Membrane Separation in Wastewater Purification and Biotechnology Applications

Dwaipayan Sen
Assistant Professor, Department of Chemical Engineering
Heritage Institute of Technology, Kolkata, India

Ankita Mazumder
Assistant Professor, Department of Chemical Engineering
Haldia Institute of Technology, Haldia, India

CRC Press
Taylor & Francis Group
Boca Raton London New York

CRC Press is an imprint of the
Taylor & Francis Group, an **informa** business
A SCIENCE PUBLISHERS BOOK

First edition published 2026
by CRC Press
2385 NW Executive Center Drive, Suite 320, Boca Raton FL 33431

and by CRC Press
4 Park Square, Milton Park, Abingdon, Oxon, OX14 4RN

CRC Press is an imprint of Taylor & Francis Group, LLC

Library of Congress Cataloging-in-Publication Data (applied for)

ISBN: 978-1-032-85984-2 (hbk)
ISBN: 978-1-032-89130-9 (pbk)
ISBN: 978-1-003-54130-1 (ebk)

DOI: 10.1201/9781003541301

Typeset in Times New Roman
by Prime Publishing Services

Preface

This book attempts to explore the applications of membrane technology in wastewater treatment and various biomedical segments. It provides an overview of different membrane technologies, starting from conventional methods like ultrafiltration, nanofiltration, reverse osmosis, to application of membrane science in hydrogel biosensors. The purpose of this book is to provide a narrative on different segments of membrane science and to create a comprehensive overview of the widespread applications of membrane for readers. This not only educates readers on membrane science, but also aims to build a framework for research among readers. In addition, there has been a constant effort to understand the technology through process modeling. For example, during the discussion of liquid membranes, two types of liquid membrane technologies have been elaborated: emulsion liquid membrane (ELM) and supported liquid membrane (SLM). Both membranes function in a similar way by transporting the desired material across the organic phase. However, their technologies differ. The first does not have any solid support, while the second has solid support, within the pores, where the organic liquid acts as a membrane. Therefore, it is evident that the transportation mechanisms will differ, which cannot be fully appreciated without understanding the underlying mathematics. Hence, a detailed mathematical elaboration is provided in the chapter. An extensive case study has been conducted using a predictive model also for a complex membrane-based technology with a high-shear membrane module. In this chapter, a fuzzy assisted artificial neural network is employed to predict permeate flux. Now the question arises: why is this necessary or what is the application of this study? The answer lies in predictive control over the membrane process. This involves predicting the permeate flux, introducing noise into the system, and understanding how the controller behaves to regain flux. This chapter might interest readers as it covers the application of artificial intelligence in membrane technology, along with the design of a controller using this AI technique.

Dwaipayan Sen
Ankita Mazumder

List of Symbols

A_m	:	Membrane surface area across which the permeation occurs
A, B	:	Verbose for the rule set (Chapter 9)
a, d, c	:	Premise parameters in the membership function set for converting verbose to a quantitative value (Chapter 9)
a	:	Interfacial area
b	:	Bias assigned to the neurons of the hidden and output layers of the ANN
c	:	Molar concentration of component
C	:	Concentration of components
C_p	:	Specific heat of fluid
C^*	:	Concentration in the vicinity of the membrane surface
C_i	:	Concentration of the solute inside the stripping phase in case of the liquid membrane
$C_{OH}^{internal}$	:	Stripping phase OH^- concentration in case of the liquid membrane
C_e	:	Concentration of the solute in the external phase in case of the liquid membrane
C_m^*	:	Concentration of the solute at the external-membrane interface in case of the liquid membrane
C_m	:	Concentration of the solute-carrier complex within the membrane phase in case of the liquid membrane
$C_{M,f0}$	:	Initial concentration of the solute within the feed in SLM and BLM
$C_{M,f}$	:	Concentration of the solute within the feed in SLM and BLM
$C_{M,m}$	:	Concentration of the solute within the membrane in SLM and BLM

$C_{M,s}$	:	Concentration of the solute within the stripping phase in SLM and BLM
D	:	Diffusion coefficient
D_e	:	Diffusion coefficient of the solute-carrier complex within the membrane phase in ELM
D_M	:	Diffusivity of the solute-carrier across the membrane phase in SLM and BLM
$D_{m,eff}$	:	Effective diffusivity of the solute-carrier within the membrane phase
D_o, D_i	:	Outer and inner diameter of the tube
d	:	Delivered oxygen
d_h	:	Hydraulic diameter of the flow channels
$d_j = d_{32}$	:	Sauter mean radius
e	:	Electronic charge
F	:	Flow rate (m^3s^{-1})
F	:	Faraday's constant (in case of electrodialysis)
final	:	A batch completion time in RDMM
h_i	:	Tube inner side fluid convective heat transfer coefficient
ΔH_v	:	Latent heat of vaporization
i	:	Current density within the electrolytic cell
J	:	Permeate flux
$\langle J \rangle$	:	Time weighted average permeate flux
k	:	Mass transfer coefficient
k	:	Reaction rate constant (Chapter 8)
k_1, k_{-1}, k_2, k_{-2}	:	Rate constant for the reaction of the metals to bind with the carrier and decomplexation
k_B	:	Boltzmann constant = 1.380649×10^{23} JK^{-1}
K_D	:	Distribution coefficient between membrane-external interface and external phase
k_1	:	First order reaction rate constant at membrane-feed interface
k_2	:	First order reaction rate constant at membrane-striping phase interface
K	:	Sorption coefficient
K_m	:	Membrane permeability
K_M	:	Michaelis-Menten constant
K_H	:	Conductive heat transfer coefficient across the tube wall

L	:	Length of the tube/channel
L	:	Radius of the hydrogel before swelling (Chapter 8)
l	:	Membrane thickness
M	:	Molecular weight of the solute
M	:	Inorganic or organic counter-ion (Chapter 7)
M_i	:	Molecular weight of the species in the distillation process
Nu	:	Nusselt number
N	:	Number of components in the multicomponent mixture
N	:	Total number of emulsion globules in case of emulsion liquid membrane
N	:	Number of datasets including training and validation data (Chapter 9)
$N_{M,m}$	:	Flux of the solute-carrier across the membrane
N_{HR}	:	Flux of hydrogen from membrane to feed side
n	:	Mole fraction of the solute
O	:	Response from the individual layer of ANFIS network
P, p	:	Trans-membrane pressure (while in case of membrane distillation it is the vapor pressure difference between feed and permeate side)
Pr	:	Prandtl number
p, q, r	:	Consequent parameters in the linear Tagaki-Sugeno type output (Chapter 9)
Q	:	Energy flux
Q	:	Mechanical transport term (Chapter 8)
q	:	Stripping distribution coefficient in the case with the ELM
r	:	Radius of the droplet on the surface
R	:	Radius of the globules in ELM
R	:	Hydrophobic group (Chapter 7)
R', R'', R'''	:	Organic group
R_{fi}	:	Inner fouling of the tube
Re	:	Reynold's number
R	:	Gas constant
R_x	:	The rate of disappearance of the solute-carrier complex in the membrane phase into the stripping phase
$R_{s,m}$	:	Sauter mean radius for membrane phase
$R_{s,s}$	:	Sauter mean radius for stripping phase

r	:	Internal radius of the membrane phase
r_s	:	Stoke's radius for the solute, in this case, the pollutant-loaded carrier
S	:	Solubility
$s = X, Y$	:	Inputs that are fed to ANFIS
t	:	Time of operation
t_{im}	:	Number of ions transported at the membrane interface
t_{is}	:	Number of ions transported at the solution vicinity to the membrane
T	:	Temperature of the fluid
U	:	Overall heat transfer coefficient
U	:	Convective source term (Chapter 8)
u	:	Velocity of the solution in the channels or swelling front velocity for hydrogel
v	:	Feed flow rate
V	:	Oxygen consumed by the tissues
Vs	:	Volumetric flow rate of a component
$V_{\text{feed phase initial}}$	:	Initial volume of the feed phase
V_{internal}	:	Volume of internal phase in a ELM
V_e	:	Volume of the external phase outside ELM
V_m	:	Volume of the membrane phase
V_i	:	Volume of the stripping phase
V_{max}	:	Maximum velocity constant
w	:	Weights assigned for the connectivity between two layers
$\overline{w}$	:	Normalized weight factor enumerated at the third layer of the ANFIS
x	:	Liquid phase mole fraction
x	:	Spatial coordinate (Chapter 8)
X	:	Length increase of the hydrogel during swelling (Chapter 8)
X	:	Halide group
y	:	Vapor phase mole fraction
z	:	Distance from the membrane surface or the direction along which the mass transfer happens
z	:	Valence number (Chapter 8)
Z	:	Electrochemical valence of ions in the solution

List of Greek Symbols used

α	:	Gas permeability coefficient
β	:	Separation factor
δ	:	Boundary layer thickness
η	:	Porosity of the support in SLM
ε_0	:	Electrical permeability
€	:	Euro currency
γ	:	Coefficient of activity
γ	:	Surface tension of the droplet on a surface (Chapter 2)
λ_n	:	Eigenvalue
μ	:	Chemical potential
μ	:	Viscosity of the feed/permeate (Chapter 2)
μ	:	Viscosity of the membrane phase in BLM (Chapter 5)
μ	:	Membership function set for converting verbose to a quantitative value (Chapter 9)
ν	:	Kinematic viscosity of the solution inside the channel
ω	:	Stirrer speed
φ	:	Property (Chapter 8)
ϕ	:	Coefficient of proportionality interrelating the flux with the chemical potential
ρ	:	Molar density
σ	:	Molar volume
θ	:	Contact angle between drop of a liquid and substrate
ζ	:	Tortuosity of the support in SLM

List of Superscripts

0	:	In contact with the membrane at the feed interface
D	:	Diffusion following Fick's law
G	:	Gas phase
G	:	Glucose (Chapter 8)
h	:	Nodes' number in the hidden layer of ANN
i	:	Nodes' number in the input layer of ANN and any layer in ANFIS
Kn	:	Knudsen diffusion
l	:	In contact with permeate interface
L	:	Liquid phase

o	:	Nodes' number in the output layer of ANN
O	:	Oxygen
sat	:	Saturated
+	:	Positive charge
–	:	Negative charge

List of Subscripts

0	:	Representing parameters at feed side
a	:	Arterial
A	:	Solute A
c	:	Pulmonary capillary
d	:	Distillate after membrane distillation
dc	:	Distillate collection
Exp	:	Experimental
f	:	Feed side of the membrane distillation unit
$feed$	:	Feed solution
fm	:	Feed side membrane surface
fb	:	Bulk feed away from the membrane surface
G	:	Glucose
i	:	Component i
$inlet$	:	Inlet to the membrane
l	:	Representing parameters at feed side
L	:	Representing parameters at permeate side
lv	:	Latent heat of vaporization
m	:	Membrane surface
M	:	Membrane
0	:	Representing parameters at permeate side
O_2	:	Oxygen
$outlet$	:	Outlet from the membrane
p	:	Permeate side
pm	:	Permeate side membrane surface
pb	:	Bulk permeate away from the membrane surface
$permeate$	:	Permeate solution
$Pred$	:	Predicted
s,i,j,k	:	Components
v	:	Venous
wi	:	Weight of component i

Contents

CHAPTER 1

Membrane Separation Technology

1.1 Introduction

The word ‘membrane’ originates from the Latin word ‘membrana,’ which means skin. A membrane is a physical barrier that selectively separates constituents from a mixture. There are several definitions of a membrane. In simple terms, it can be defined as a thin barrier placed between two distinct mediums or phases, allowing the selective permeation of one or more components of feed stream from one medium to the other. Transport across a permselective membrane is primarily induced by a driving force such as pressure gradient, concentration gradient, temperature gradient, or electric field gradient across the membrane. At the microscopic level, any membrane separation process is an amalgam of both momentum and mass transfer.

Evolution of membrane technology has experienced a long history in lab scale studies prior to accomplishing its first promising industrial application in the 1960s. In the initial years of evolution, membrane separation processes were mainly used as analytical tools in biomedical and chemical laboratories. However, application of membrane technology has now been boosted in wide domains which include beverage, food, medical, pharmaceutical, chemical process industry, wastewater treatment facilities, and desalination processes. Its widespread popularity is due to its advantages over other existing separation processes such as distillation, adsorption, extraction, and crystallization. Some notable benefits of membrane separation include low capital investment, minimal footprint area requirement, reduced energy demand, superior separation efficacy, easy integration with other separation processes, straightforward scalability, environmental friendliness, and non-requirement of secondary separation methods. The membrane separation efficiency is dependent on various factors including membrane porosity, distribution of membrane pore size, pore diameter, size and dimensions of the solutes, the affinity in between the membrane material and feed stream, and diffusivity/solubility. There are wide varieties of fabrication methods for synthesis of membranes by utilizing different types of materials. In spite of these benefits of membrane processes and various available options for fabrication and applications of membrane, there is enough scope yet for more advancement.

1.2 Classification of Membrane Separation Processes

Several industries including pharmaceutical, chemical, biotechnology, oil, food industries, and effluent treatment plants implement membrane based technologies either for purification or concentrating the feed through separation. Membrane separation processes may be categorized on the basis of different aspects including the nature of driving forces, configurations, types, and transport mechanisms. Based on the existing driving forces and their separation characteristics, these membrane separation processes are used for varied operations. Table 1.1 highlights the purview of membrane processes classified depending on the varied driving forces (including pressure-driven, thermal-driven, osmosis-driven, and electrical-driven).

Pressure-driven membrane based processes are further categorized based on the pore size of the membrane into the following classes: microfiltration (MF), ultrafiltration (UF), nanofiltration (NF), and reverse osmosis (RO). These processes are specified in a descending order of membrane pore sizes, with microfiltration (MF) having the largest range of pores size and reverse osmosis (RO) having the smallest, as indicated in Table 1.1. In these pressure-driven processes, separation occurs via the size exclusion mechanism. Additionally, there are many other promising membrane separation processes such as dialysis, pervaporation, gas separation, electrodialysis, and membrane distillation. These processes are driven by various forces, including concentration gradients, electrical potential gradients, or temperature gradients. These processes are well-recognized and utilized in industries for various applications including desalination of saline water, hemodialysis, and the separation of biological species, organics, metal ions, and azeotropic mixtures etc. For instance, porous membranes are used in electrodialysis and dialysis, whereas nonporous or dense membranes are preferred for pervaporation. In this chapter, we will discuss pressure-driven and concentration-driven membrane separation processes in detail.

Depending on the mode of operation, the membrane based processes may be further categorized into batch, continuous, or fed batch processes. Besides that, the processes may also be classified into cross-flow filtration and dead-end filtration based on the direction of feed introduction to the membrane module. In case of the batch recycling of permeate, dead-end filtration is usually adopted. Again, the membrane process is considered to be fed batch when a substantial amount of the feed components are yet present in the permeate stream. In case of the continuous processes, the feed is being continuously fed to the membrane. During dead-end filtration, the permeation of feed takes place across the membrane along with the continuous accumulation of the retentate over the membrane surface. This gradually results in permeate flux reduction on account of the fouling of membrane by the retentate accumulation on the surface. However, with the application of cross-flow filtration method, the feed is flown in a direction

Table 1.1: Some of the common membrane processes categorized based on driving force

Membrane process	Driving force	Mechanism of separation	Pore size range	Types of membrane used
Microfiltration (MF)	Pressure gradient	Sieving mechanism	100–10000 nm	Both asymmetric and symmetric microporous membrane
Ultrafiltration (UF)	Pressure gradient	Sieving mechanism	10–100 nm	Asymmetric microporous membrane
Nanofiltration (NF)	Pressure gradient	Solution diffusion mechanism	0.5–5 nm	Asymmetric skin type membrane
Reverse osmosis (RO)	Pressure gradient	Solution diffusion mechanism	<1 nm	Asymmetric skin type membrane
Dialysis	Concentration gradient	Diffusion mechanism	<1 nm	Thin membrane
Pervaporation	Concentration gradient	Solution diffusion mechanism	<1 nm	Asymmetric microporous membrane
Membrane distillation	Temperature gradient	Temperature gradient	1–10 nm	Microporous membrane
Electrodialysis	Electric potential gradient	Electric potential gradient	<1 nm	Charged membrane

parallel to the membrane surface and consequently the feed stream gets separated into permeate and retentate streams. The cross-flow system is aided with the benefit of less fouling because the retentate is continuously dissipated during the operation resulting in no accumulation of retentate on the membrane surface. The dead-end process is typically preferred for batch operations, where membrane cleaning can be performed after each run. Conversely, cross-flow filtration is suitable for continuous operation of membrane processes, as membrane fouling is less of an issue.

1.2.1 Microfiltration (MF)

Microfiltration (MF) is a pressure-driven membrane separation process capable of retaining particles or biological components within the micron size range (0.1–10 μm) and molecular weight (MW) greater than 100 kDa. The retention abilities depend on the size exclusion mechanism and pore size of microporous membranes used in MF. In MF, the pore sizes of the microporous membrane are relatively large ranging between 100 nm and 10000 nm. As the membrane pore size is quite large, the pressure required to execute separation is comparatively low ranging between 1 bar to 3 bar. While the separation of micron-sized particles can be achieved using depth materials or non-membrane mediums, only a membrane with precisely specified pore sizes ensures quantitative retention. Depth filters are commonly utilized for clarifying operations which do not require quantitative retention. On the contrary, membranes are favored for filtration or pre-filtration purposes to extend the lifetime of other downstream membranes. Commonly, MF technology is employed for the separation of sediments, pigments, paints, suspended solids, algae, bacteria, protozoa, and yeast cells. However, the separation of microscopic particles such as colloids, monovalent species (sodium (Na^+) ions, chloride (Cl^-) ions), viruses, and dissolved organic matter is not feasible by using MF. In the case of a MF system, the configuration of the membrane can be spiral wound, flat sheet, hollow fibres, hollow fine fibre, or tubular. The feed velocity in a tubular membrane (cross-flow) or in a flat sheet membrane (dead-end) is generally maintained at 1 to 3 m/s at low or moderate pressure. MF is mostly used in combination with UF or RO, as can be seen in desalination of wastewater treatment plants. MF also plays a vital role in the dairy, biotechnology, pharmaceutical and beverage industries. The benefits and shortcomings of MF are discussed in Table 1.2.

Table 1.2: Benefits and shortcomings of pressure-driven membrane processes

Membrane separation process	Benefits	Shortcomings
Microfiltration	Low operating pressure	The separation of biological particles and suspended matter is only possible
	Lower energy demand compared to other membrane separation processes	Highly sensitive to oxidizing agents
	Easy and uncomplicated maintenance	Membrane damage can occur if the applied cleaning pressure exceeds 100 kPa
	Fouling is not intense due to the large membrane pore size and the utilization of low pressure	Likelihood of membrane damage from sharp or hard particles during membrane operation
	Cost effective	
Ultrafiltration	It can be utilized to separate a wide range of feed components due to its preferred pore size range	Ultrafiltration (UF) cannot separate dissolved salts or low molecular weight species. Therefore, it is not suitable for desalination of saline water.

Contd...

Contd...

	Low energy demand separation occurs without any phase change	Efficacy of UF is not satisfactory for separation of macromolecular mixtures. It is mainly effective if the difference in molecular weight of components in the mixture is at least 10 times or higher.
	Operation at low to medium pressure	
	Useful membrane technique for separation of temperature-sensitive components used in food, pharmaceutical, biotechnology industries	
	Easy handling and operation because of its simple and compact nature	
Nanofiltration	Lower operating pressure than RO	Concentration polarization
	Operating expense is lower because it is performed at low pressure and room temperature	High maintenance expenses.
	Ability to continuously process large volumes	Short longevity of membrane
	Addition of extra ions for water softening is not required as in the case of ion exchangers	
Reverse Osmosis	Desalination of brackish and sea water can be performed	Requirement of high pressure
	Separation without phase change	Pretreatment of the feed is essential
	Compact module. So, less space is consumed in comparison to other desalting systems	Lower permeate flux rate
	Easy and uncomplicated maintenance	Energy-intensive process
	Easy scaling up	Intense fouling

1.2.2 Ultrafiltration

Ultrafiltration (UF) is another pressure-driven filtration process with nominal pore size ranging between 1 to 100 nm. In this process, hydrostatic pressure induces the transportation of the liquid through a semi-permeable membrane, where the small-sized particles, suspended solids, dissolved molecules, etc. present in the feed stream are retained on the basis of their molecular size. Here, the separation of components occurs based on the size-exclusion based sieving mechanism in which the wider and large sized components are retained on the UF membrane surface while narrower and smaller components will permeate through the membrane. The permeability through the UF membranes may be dependent on the molecular, chemical, or electrostatic characteristics of the feedstock. The retention ability is also correlated to the membrane's MWCO. Particles of molecular weight in the range 1–1000 kDa are generally retained by UF membrane. However, water and salts pass through the membrane in the permeate. From the fundamental point of view, UF is not distinct from other pressure-driven membrane processes such as MF, NF, or gas separation, except in terms of the retention ability of membrane depending on the size of particles. Lower operating pressure is necessary for UF operation as compared to other pressure-driven membrane separation processes including NF and RO. This is attributed to the porous characteristics of the UF membranes having larger size pores than these above-mentioned membrane processes. Thus, high permeation flux can be obtained from UF operation being executed at a pressure of 200 to 1000 kPa.

In the initial stages of its development, ultrafiltration (UF) was typically explored on a laboratory scale for the purification and separation of various compounds. The initial UF membranes were not suitable for industrial commercial applications, as they lacked sufficient mechanical and chemical strength and properties. Then, in the 1960s, the fabrication of asymmetric membranes by Loeb and Sourirajan paved a new route for the development and utilization of efficient UF membranes on a commercial and industrial scale (1). These asymmetric membranes were both selective and permeable. This provided a much-needed impetus to the membrane science research communities, enabling the development of commercially successful membranes. In the late 1960s, the UF-based industrial plant was installed, where it was used for fractionation technology. Since then, UF technology has gradually evolved into an advanced separation technique, finding applications in numerous industrial sectors. Some of the notable industrial applications include water treatment, wastewater treatment and reclamation, resource recovery, chemical recovery, dairy production, cell harvesting, medical use. UF is often preferred over traditional disinfection and water treatment processes due to its easy and uncomplicated operation, lower energy usage, reduced chemical demand, limited control methods, and high treatment efficacy. In terms of energy consumption, UF is not a cost-intensive operation, as the driving force (pressure gradient) is quite small. Therefore, the substantial expenses are related to the membrane module and membrane replacement. Table 1.2 highlights the benefits and shortcomings of the UF process.

Today, several large-scale UF plants are successfully operating in various countries including the USA, Singapore, Belgium, Netherlands, Spain, Germany,

Kuwait, and Namibia. The market size of UF membranes is steadily elevating with the continuously growing awareness on environmental pollution, reduction of freshwater resources, sustainability policies, and stringent government regulations. In 2017, the market was around USD 950.0 million, which is expected to reach USD 2,140.1 million by the year of 2023. The Koch Membrane System, Dow Chemicals, Oasys Water, PennWell Corporation, and GE Corporation are the top five companies dominating the global UF market. Among them, Dow Chemical and Koch Membrane System hold more than 51% of the global market share. Moreover, the U.S.A. and Asia Pacific regions dominate the global membrane market,accounting for around 65% of the share due to the increasing demand in wastewater management, chemical processing, and pharmaceutical sectors.

1.2.3 Nanofiltration (NF)

Nanofiltration (NF) is a comparatively recently developed pressure-driven membrane separation process. Considering its attributes, NF is a molecular-level separation process that exists in a transition regime between the size exclusion mechanism dependent UF process and thc solution–diffusion model-based RO operation. Due toits reduced energy requirements compared to RO and its higher rejection capacity than UF, NF-based processes are often preferred in several applications. This separation technique is also receiving significant attention in the field of wastewater treatment. The nominal membrane pore size of NF ranges from 0.5 nm to 5 nm. NF operation is usually executed at a pressure range of 0.5–1.0 MPa. Due to the very small pore size of NF membranes, they can retain even minute-sized particles, divalent ions, multivalent ions, and uncharged particles. However, monovalent ions easily permeate through the membranes because of their surface electrostatic nature. These characteristics favor NF technology for fractionation purposes and the selective retention of solutes from a mixed process feed stream. In the NF separation process, both particle dimension and electrostatic charge play a pivotal role, as NF possesses properties of both RO and UF. NF membranes are also referred to as charged UF membranes or low-pressure RO membranes in view of UF and RO, respectively. Henceforth, the sieving mechanism will be prominent for the separation of large sized particles and colloids, while solution diffusion and charge effect will be prominent for retention of small sized particles and ions. It is preferred for several applications in textiles, paper and pulp, dairy, metal recovery, and pharmaceutical industries specifically for the removal of particles of dimension around 1 nm.

1.2.4 Reverse Osmosis (RO)

In simple terms, reverse osmosis (RO) operation is the opposite of the osmosis process. In osmosis, a semi-permeable barrier is placed between two solutions of different concentrations, and the solvent flows across the barrier from the side of lower solute concentration to the side of higher concentration. On the other hand, if the solvent flows across a semipermeable membrane from the side of higher solute concentration to the side of lower solute concentration due to an external force, this process is called reverse osmosis (RO). The driving force behind the normal osmosis process is the reduction in the free energy of the system, which gradually reduces as the system approaches equilibrium. After equilibrium is reached, no further osmosis occurs. However, in the process of RO, the applied external force, which is larger than the system's osmotic pressure, becomes the driving force. Thus, RO based membrane separation is also a pressure-driven membrane process. The difference with other pressure driven membrane separation processes is that they primarily follow a size exclusion mechanism for separation, whereas RO utilizes diffusion phenomenon.

Although osmosis was discovered by Jean-Antoine Nollet in 1748, it was only known as a natural phenomenon without any experimental demonstration (2). In 1950, the desalination of seawater using a semi-permeable membrane was first investigated on a laboratory scale, but the process was not commercially feasible due to the very low flux rate. In this context, the development of asymmetric membranes by Loeb and Sourirajan in 1962 made the RO operations viable on a commercial scale, as these membranes were able to reject salts with comparatively higher flux rates (1). Later in subsequent years, John Cadotte of the FilmTec Corporation patented the interfacial polymerization process of trimesoyl chloride and m-phenylene diamene to synthesize asymmetric membranes with much higher flux rates and lower salt permeation (3). This synthesis procedure was widely accepted for the fabrication of efficient and commercial thin RO membranes.

Generally RO membranes are classified into two main categories: asymmetric membranes and thin film composite membranes. In case of the asymmetric RO membranes, only one polymer material is used, which contains a thin permselective skin layer supported by a higher porous sub-layer. Here, the dense skin layer (0.1–1 μm) plays a key role in determining the flux rate and selectivity, while the supporting porous sub-layer has an insignificant contribution to the membrane separation characteristics. Fabrication of asymmetric membranes is commonly done using the phase inversion or polymer precipitation technique, as developed by Loeb and Sourirajan. Substantial advancements were achieved in the 20 years following the initial development of asymmetric RO membranes made of cellulose acetate. Cellulosic polymeric membranes made from cellulose acetate, cellulose triacetate, and linear aromatic polyamide membranes are among the widely used asymmetric RO membranes. Moreover, another category of thin-film composite RO membrane is synthesized from either two or more polymer materials and consists of a thin polymer barrier layer supported on one or more porous support layers. These support layers are structurally distinct from the top surface layer. The surface (top) layer is extremely thin, measuring 0.1 μm or less, and provides a higher flux rate. The most widely used thin film composite RO membranes are synthesized from cross-linked aromatic polyamide, supported by a layer made of polymers like polysulfone or aryl-alkyl polyetherurea, using the interfacial polymerization method. The water permeates across the membrane by following a tortuous path. The RO process is able to retain the smallest species, including monovalent ions, viruses, and dissolved organic content, from the feed components that other membrane based processes fail to separate. Although the RO process can be implemented in both cross-flow and dead-end configurations, cross-flow mode is usually preferred because of low energy requirements and less fouling propensity.

RO has widespread applications in almost all industries, including chemical processing, metal-finishing, mining, dairy, printing, food processing industry, pharmaceutical, textile, paper, and wastewater treatment. Among these, desalination is the most widely utilized. For example, in the metal finishing industry, RO is utilized for recovery of metals and the reuse of plating rinse water. Additionally, the production of miniature electronic components (semiconductors) requires water of very high purity. Therefore, RO based systems are used to purify bulk volumes of spent ultrapure water. Similarly, in the pharmaceutical industry and for medical usage, RO membranes are used to provide ultrapure water. RO technology is also utilized in the printing and photographic industries to mitigate their discharge issues along with the benefit of deducing water expense through recovery. Furthermore, RO based separation processes may be used in conjugation with other separation technologies like UF, distillation, and pervaporation to provide selective and efficient separation.

1.2.5 Gas Separation through Dense Membranes

Gas separation is one of the promising applications of membrane separation technology. Polymeric membranes are usually preferred for gas separation operations. Generally, gas shows a tendency of solubility inside a polymeric membrane, which is greatly dependent on the type of used polymer in the membrane and the characteristics of gas. Three major steps are followed in a gas separation process which include sorption, diffusion, and desorption of gas. In the first step of sorption, the dissolution of a gas takes place into the polymer by which a membrane is made. The second step is diffusion, in which the gas from the feed side is permeated across the membrane to the permeate side. On the feed side of the membrane, the partial pressure is higher than on the permeate side. Then the final step is desorption of the gas on the permeate side. This membrane gas separation follows the solution–diffusion mechanism where the driving force responsible for separation is both concentration gradient and partial pressure gradient of the gas across the two membrane sides. This solution-diffusion mechanism is explicitly discussed in the successive sections.

In the 1980s, membrane based gas separation technology had made a key breakthrough. It was the development of revolutionary composite membranes by Henis and Tripodi through coating a thin layer of silicone-rubber on the asymmetric substrate for achieving higher carbon monoxide/hydrogen selectivity in a less expensive module system. This resulted in the commercialization of the first large-scale gas separation based membrane system (Prism1) by the Monsanto Co. in 1980, for the recovery of hydrogen gas from the purge gas released from the ammonia plant. Monsanto's success in gas separation has induced extensive research efforts in the development of novel membranes and separation processes. In the following years, Dow developed gas separation technologies for separating nitrogen from air. Also, Cynara and Separex performed the separation of carbon dioxide from natural gas. In 1990, defect-free ultra-thin asymmetric membranes were fabricated by adopting a dry/wet phase inversion method. for enhanced gas permeation across the membrane, satisfying the growing demand in large-scale industrial operations. Subsequently, rigorous efforts were made to synthesize highly efficient asymmetric hollow fiber membranes for better and more selective air permeation. The setup of a simple membrane-based gas separation process is not much different from that used for solid–liquid separation, except for the presence of different phases in the streams. A gas mixture is introduced at very high pressure in the membrane module, where one component diffuses rapidly through the membrane and gets enriched in the permeate stream, while the rest of the gas becomes concentrated in the retentate stream. To ensure higher and complete separation, recycling a fraction of the permeate stream or residue stream may be essential.

1.2.6 Dialysis

Dialysis is a membrane based separation process in which the concentration gradient acts as the driving force for the separation of solute and solvent. Due to the concentration gradient, the solute diffuses and then passes through the membrane. The solutes have varied diffusion rates, depending on the differences in their molecular size and solubility. With the development of the Cerini dialyzer in Italy, dialysis was recognized as the first membrane separation process to be implemented on an industrial scale. In the 1930s, the production of rayon from cellulose gained noteworthy industrial expansion, leading to a crucial requirement to recover sodium hydroxide from the by-product stream of the process containing hemicellulose/sodium hydroxide solution. In this context, the dialysis method was adopted using a fine microporous membrane for the separation of the concentrated hemicellulose solution from water. The smaller sized sodium hydroxide molecules diffuse through the membrane due to the concentration gradient, delivering an uncontaminated product stream. Because of its cost-effectiveness, it was widely used until the development of pressure-driven processes, including UF and MF in the 1960s to 1970s. However, after that period, applications of dialysis in industries largely diminished as it is a diffusion-dependent separation process. Diffusion is a relatively slow phenomenon and less selective compared to other newly developed technologies. Processes like UF and electrodialysis are preferred due to rapid separation and their high selectivity, achieved by various better selective membranes. In the dialysis process, it is preferred to use a thin membrane to obtain a higher flux rate.

The dialysis process is able to conduct specialized applications, particularly in the fields of biotechnology and life sciences. Currently, two very important applications of dialysis have been developed. These applications include the dialysis of blood, known as haemodialysis in the artificial kidney, and the exchange of oxygen and carbon dioxide in blood in the artificial lung. Both processes require low pressure and mild dialysis conditions. Unlike other processes, separation of sensitive substances can be completed using dialysis without any damage. This is because the dialysis process is generally conducted under mild environmental conditions, such as ambient temperature, no substantial transmembrane pressure decline, and low shear flow rate. Although slower compared to other pressure-driven processes, dialysis can reliably distinguish small sized molecules from large ones. This is because the absence of pressure difference across the membrane limits the convective flow through defects in the membrane. This benefit is desirable in critical applications, including medical and immunological separations and the removal of salt from genetically engineered protein solutions. In such applications, the leakage of any unwanted species from the feed solution to the permeate is not permissible.

1.3 Overview on Transport Mechanism

The mathematical explanation of the membrane permeation mechanism is implicitly based on the thermodynamic branch, which deals with different driving forces, including pressure gradient, concentration gradient, temperature gradient, and electrical gradient. All these gradients are correlated with each other. So, the overall driving force responsible for the permeation of a component *i* is the chemical potential (μ_i). So, the flux J_i (kg/m^{2}s) of component *i* is expressed as:

$$J_i = -\phi_i \frac{d\mu_i}{dx} \tag{1.1}$$

where, $d\mu_i/dx$ represents the gradient of chemical potential of component *i*, and ϕ_i is the coefficient of proportionality interrelating the flux with the chemical potential. The different driving forces mentioned above can be reduced to the gradient of chemical potential and can be further described in terms of this given equation. This analogy is quite beneficial in cases where some membrane separation processes encounter more than one driving force. For instance, reverse osmosis, gas separation, and pervaporation involve both pressure gradient and concentration gradient as the driving forces to accomplish the membrane separation. Assuming the driving forces for a membrane process are restricted to only concentration gradient and pressure gradient, the chemical potential is then expressed in the following manner:

$$d\mu_i = RTd\ln(\gamma_i n_i) + \sigma_i dp \tag{1.2}$$

where, n_i signifies the mole fraction of component *i*, γ is the coefficient of activity, *p* is the pressure, and σ_i is the molar volume of the component *i*.

In case of incompressible phases (solid and liquid membranes), there is no significant variation in volume with respect to pressure. So, after integrating equation 1.2 with respect to pressure and concentration

$$\mu_i = \mu_i^0 + RTd\ln(\gamma_i n_i) + \sigma_i (p - p_i^0) \tag{1.3}$$

where μ_i^0 is the chemical potential for pure component i at a reference pressure, p_i^0.

However, in case of compressible phases (gas), the pressure imparts a significant effect on the molar volume of the gases. Substituting the gas laws and integrating equation 1.2, the equation is deduced to

$$\mu_i = \mu_i^0 + RTd\ln(\gamma_i n_i) + RT\ln\left(\frac{p}{p_i^{sat}}\right) \tag{1.4}$$

Permeation through the membrane is usually interpreted based on the two models: the solution–diffusion model and the pore flow model. Both the models are schematically represented in Figure 1.1. In the case of the solution–diffusion model, the feed is initially dissolved within the membrane material and subsequently permeation of feed occurs across the membrane. This permeation occurs as a result of the concentration gradient, which performs as the main driving force for this separation process. Henceforth, the efficiency of membrane separation is regulated by the degree of solubility and the permeation rate of the feed components. On the contrary, in the pore flow model, permeation of the feed components takes place across the membrane pores because of the pressure gradient across the feed and permeate side of the membrane. Here, the pressure gradient acts as the prime driving force for this separation process. The flow of feed components across the pores of the membrane is convective in nature. The separation of permeate occurs depending on the size exclusion concept, where the components having size less than the pores get permeated whereas the components of greater size than the pores are restricted from passing.

For defining any permeation model, some assumptions are to be considered. In the case of membrane transport, two prime assumptions are considered. The first assumption is that an equilibrium exists between the phases present on both the permeate and retentate sides of a membrane. This signifies the presence of a continuous gradient in chemical potential through the membrane. It must be implicitly understood that the rate of transport of the feed components is more at the membrane interface in comparison to the diffusion of feed components across the membrane. This is valid for every membrane separation process except the few transport cases involving chemical reactions (for example, facilitated transport or gas diffusion in metals). In such exceptional cases, the rate of transport of the feed components at membrane interfaces is generally slow. Secondly it is assumed that the applied pressure to any membrane process is uniform throughout the system.

1.3.1 Solution–diffusion Model

The solution–diffusion model depends on the diffusion mechanism and concentration gradient. Both diffusion and concentration are closely related, molecules are transported from one place to the other based on the concentration gradient in the system. Henceforth, the molecules are transported in the direction of the concentration gradient during the diffusion phenomenon. This concept was first introduced theoretically and validated practically by the scientist Fick in 1857.

Therefore, according to the solution–diffusion model, the pressure is uniform in the system, and the concentration gradient alone governs the chemical potential. Taking this assumption into account, equations 1.1 and 1.2 are combined and may be rewritten as shown in equation 1.5 (considering σ_i as constant)

$$J_i = -\frac{\phi_i RT}{n_i}\frac{dn_i}{dx} \tag{1.5}$$

The gradient of component i in equation 1.5 may be expressed in terms of the gradient of the mole fraction of component i in a more significant manner utilizing the concentration term c_i (kg/m^3).

$$c_i = M_{wi}\rho n_i \tag{1.6}$$

where M_{wi} is the molecular weight of the component i (kg/mol) and ρ is the molar density (mol/m^3). Thus, equation 1.5 is rewritten as

$$J_i = -\frac{\phi_i RT}{c_i}\frac{dc_i}{dx} \tag{1.7}$$

The expression of equation 1.7 is resembles Fick's law, where the term $\frac{\phi_i RT}{c_i}$ is analogous to the diffusion coefficient D_i in equation 1.8,

$$J_i = -D_i\frac{dc_i}{dx} \tag{1.8}$$

where J_i is the transfer rate of component i in the feed across the membrane (kg/m^2.s), $\frac{dc_i}{dx}$ is the concentration gradient of component i, and D_i is the diffusion coefficient (m^2/s) which signifies the extent of the mobility of the molecules of an individual component i. Here, the negative sign represents the direction of diffusion along the driving force of the concentration gradient.

Now, integrating equation 1.8 within the limits of membrane thickness (l) provides

$$J_i = \frac{D_i\left(c_i^0 - c_i^l\right)}{l} \tag{1.9}$$

where c_i^0 and c_i^l are the concentrations of component i in contact with the membrane at the feed interface and the permeate interface, respectively.

1.3.2 Pore Flow Model

The pore flow model depends on the pressure-gradient as the driving force for the convective flow of the feed components across the pores of the membrane. In 1856, Henry Philibert Gaspard Darcy introduced Darcy's law based on his experimental data. This law essentially governs fluid flow through any porous medium, which is known as the pore flow model. The expression of Darcy's law is given below.

$$J_i = kc_i \frac{dP}{dx} \tag{1.10}$$

where $\frac{dP}{dx}$ denotes the pressure gradient across the membrane sides, and k is an intrinsic property of the membrane representing its permeability coefficient. The elementary diffusion phenomenon is comparatively much slower than the convective fluid flows through the membrane driven by pressure-gradient.

The solution–diffusion model is generally followed in the case of permeation through dense or nonporous membranes, as the membrane pores are very small. Indeed, the pores in dense membranes are precisely the miniscule spaces existing or arising in between the polymer chains due to the thermal movement of the molecules. These spaces are continuously appearing and disappearing with the course of permeation of the feed components across the membrane. Contrarily, porous membranes usually follow the pore flow model governed by Darcy's law, as the pores in these non-porous membranes are fixed, interconnected, and prominent in nature. The characteristics of the pore flow model largely depend on the size of the membrane pores.

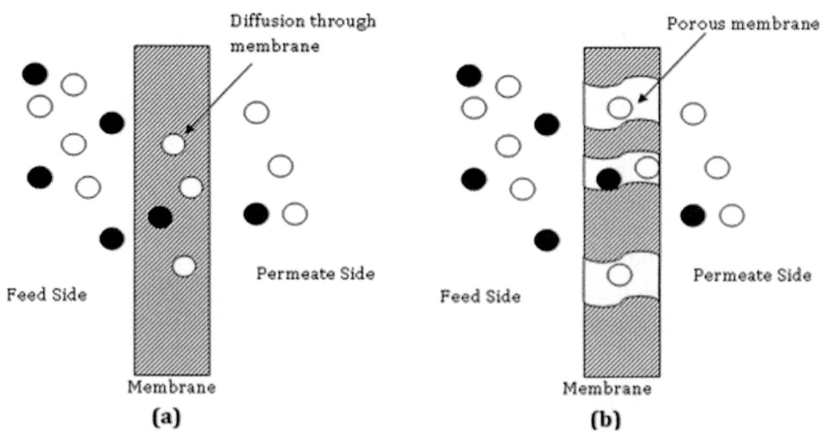

Fig. 1.1: Schematic representation of (a) solution-diffusion model; (b) pore flow model

1.4 Membrane Modules

Membranes with a large surface area are necessary for better separation in commercial or industrial scale operations. Therefore, membranes must be packed or arranged in units that are both compact and spacious to fulfill the area requirements for separation. Membrane modules are assemblies of these membranes. In the most basic membrane module, the feed is introduced through the feed inlet at the specified operating conditions (pressure, temperature, and flow rate). Then the feed stream gets separated into retentate and permeate streams based on the membrane characteristics, and the feed composition. There are different module configurations for flat sheet/disk and tubular/hollow membranes. Among them, plate and frame and spiral wound membranes modules are used to assemble flat sheet membranes. Moreover, tubular (approx membrane diameter of less than 10 mm), capillary (approx membrane diameter in the range of 0.5–10 mm), and hollow fiber (approx membrane diameter of less than 0.5 mm) modules are employed for tubular or hollow membranes.

According to the process demand, membrane separation processes use one or more modules among varied designs in distinct configurations. The type of membrane module, its configuration, and the entire system design depend on the engineering parameters, with the foremost objective of attaining maximum separation efficiency with a cost-effective design or membrane separation process. The key factors include feed composition and its characteristics, compactness, scale, operation, maintenance, cleaning, and fouled membrane replacement. In the subsequent sections, varied types of membrane modules for both flat sheet and tubular membranes will be elaborated upon, along with their design and working principles. A comparison between the different types of membrane modules is represented in Table 1.3.

1.4.1 Plate-and-Frame Module

The Plate-and-Frame module is the simplest module for assembling flat sheet membranes in the form of a casket. A schematic representation of the Plate-and-Frame module is provided in Figure 1.2. In this module, membranes are arranged in a stacked manner, with one membrane placed on top of one. The feed side and permeate side of the membranes face each other, creating individual compartments that consist of a set of membranes. Spacers are fitted between these compartments. All the sets of membranes or compartments in this type of module are sealed using closing rings. Finally, the module is embedded between plates to develop a Plate-and-Frame stack module. The feed stream is fed into the module from the feed side and flows adjacent to the membrane surface, where a fraction of the feed stream through it to the permeate channel and ultimately gets collected in a central permeate collector. The packing density in the range of 100–400 m^2/m^3. Usually, the plate-and-frame module is adopted for small-scale applications because of its low packing density, expensiveness,

Table 1.3: Comparison of different membrane modules

Type of membrane module	Plate and frame module	Spiral wound module	Tubular module	Capillary module	Hollow fiber module
Packing density (m^2/m^3)	100-400	300-1000 m^2/m^3	300 m^2/m^3	600- 1200 m^2/m^3	Maximum upto 30,000 m^2/m^3
Channel spacing (cm)	0.03-0.25	0.03-0.1	1-2.5		0.02-0.25
Manufacturing cost	High	Moderate	High	Moderate	Low
Energy cost	Moderate	Low	High	Low	Low
Fouling control or ease of cleaning	Good	Poor-fair	Excellent	Good	Poor
Pressure drop in permeate side	Low	Moderate	Low	Moderate	High
High pressure operation	Marginal	Suitable	Marginal	Not suitable	Suitable
Membrane processes	Pervaporation and electrodialysis	RO, UF, gas separation	UF	UF	RO, gas separation
Common applications	Not commonly used nowadays. Used mainly in pervaporation and electrodialysis.	Brackish water and industrial effluent treatment; Highly contaminated feed gas treatment where intense concentration polarization is a problem, and high flux is required.	High fouling feeds.	Effluent treatment.	Desalination; High volume applications with lower flux and less concentration polarization in case of gas separation (nitrogen separation from air).

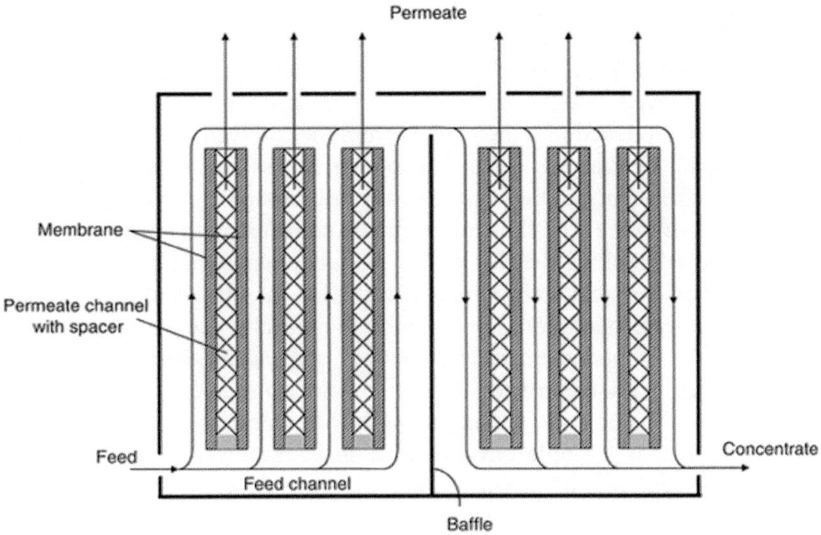

Fig. 1.2: Schematic representation of Plate-and-Frame module (4)

and external and internal sealing issues. The use of the plate-and-frame module is very limited nowadays. Only some membrane separation processes involving pervaporation and electrodialysis use this configuration module in large numbers. Commonly, spiral-wound modules are preferred. In the case of ultrafiltration (UF) and reverse osmosis (RO), plate-and-frame modules are mostly employed for handling feed with high viscosity or intense fouling tendency.

1.4.2 Spiral-wound Module

The spiral-wound module is similar to plate-and-frame modules, with the exception that flat sheet membranes separated by spacers are spirally wound around a central permeate collection tube, as shown in Figure 1.3. This module consists of flat membrane sheets, spacers, and a common porous permeate collecting tube. The membranes form an envelope shape, with permeate side

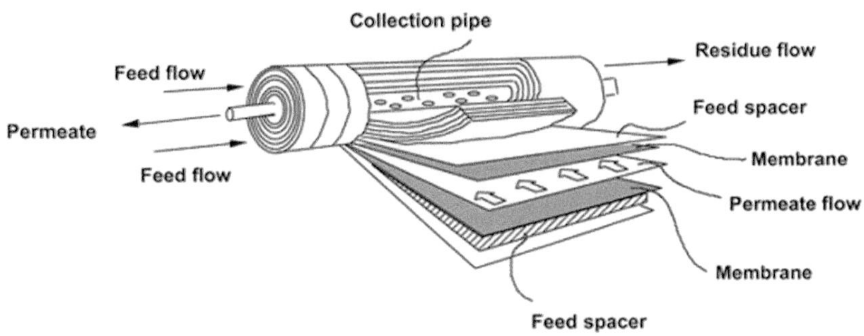

Fig. 1.3: Schematic representation of spiral-wound module (5)

spacers positioned between two consecutive membranes. This envelope is closed on three sides, with the remaining fourth side open and attached to the central permeate collection tube. This membrane configuration is currently the most commonly employed module for flat sheet membranes. In this configuration, the feed stream is introduced into the module from the end wall and flows in parallel direction to the central tube, passing axially across the membrane module. Concurrently, the permeate flows radially inward through spacers in a cross-flow direction relative to the feed flow and is finally collected in the central collection tube. The normal packing density of this module is approximately in the range of 300–1000 m^2/m^3, relatively larger than that of the plate-and-frame module. For industrial scale operations to be both efficient and economically feasible, numerous spiral wound membrane envelopes are packed together in a single pressure vessel. With the use of a multi-envelope module, the pressure drop experienced by the permeate passing towards the central pipe can be minimized to a large extent. If a single envelope design is employed to provide a large membrane surface area, a significant pressure drop develops across the central collection pipe due to the long path length that the permeate has to travel. The length of the module or central permeate collection tube depends on the module diameter. Usually, spiral wound modules with dimensions of 1 m length and 0.2 m in diameter are utilized in industrial operations.

1.4.3 Tubular Module

Unlike capillaries and hollow fibers, tubular membranes are not self-supporting. Henceforth, these tubular membranes are placed in a porous tube made of metal, polymer, or ceramic materials. Inside this tube, there is no limitation on the number of tubular membranes that can be placed, and it can contain numerous membranes. The pictorial representation of the tubular membrane module is provided in Figure 1.4. In this module, the introduced feed stream flows through

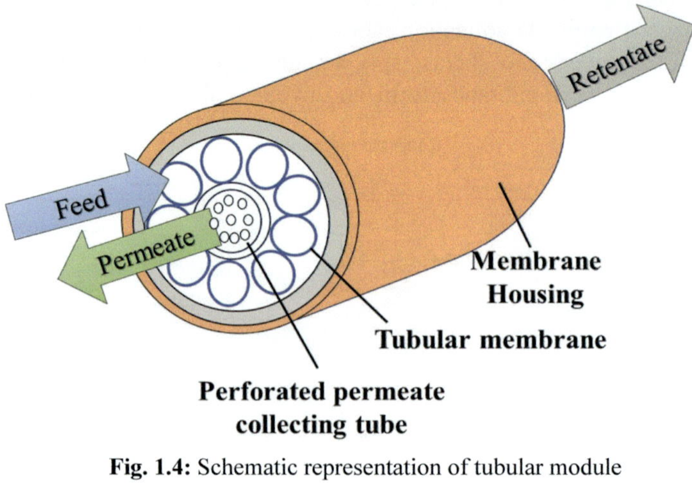

Fig. 1.4: Schematic representation of tubular module

the tubular membranes packed in the tube, while the permeate is collected through the supporting porous tube. In most cases, these tubular modules are employed to accommodate ceramic membranes with a normal packing density of 300 m^2/m^3.

1.4.4 Capillary Module

A large number of capillaries are placed together inside a capillary module. As the capillaries are self-supporting in nature, housing is not required to assemble these capillaries as a membrane module. Figure 1.5 represents the schematic of a capillary membrane module. During the compilation of the capillaries inside the module, the ends are sealed using agents like polyurethanes, epoxy resins, or silicone rubber. The operation of this module can be executed in two different ways. In the first case, the feed stream is passed via the lumens of the capillaries and the permeate flows outside the capillaries. However, in the second case, the feed flows outside the capillaries, while the permeate passes through the lumen of capillaries. The method chosen for operating a capillary module depends on various process parameters, including transmembrane pressure, pressure drop across the membrane, and membrane material. For instance, in a porous membrane, the capillaries have a gradient in pore size across them. Therefore,the mode of operation is selected based on the position of the smallest pores on the capillaries, whether they are outside or inside. A capillary module has a nominal packing density of 600–1200 m^2/m^3.

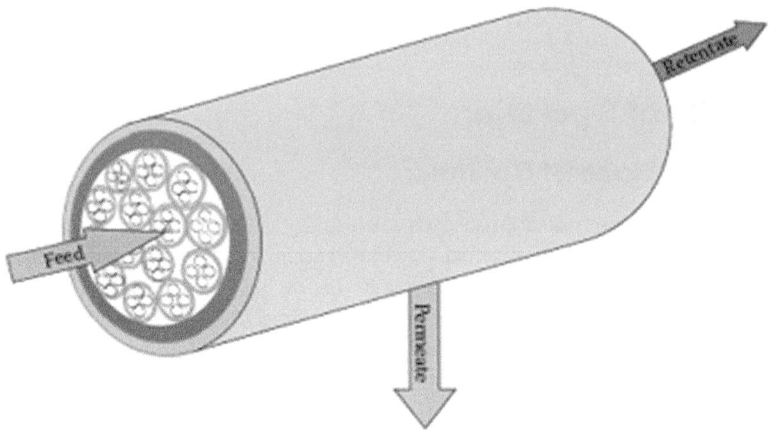

Fig. 1.5: Schematic representation of capillary module

1.4.5 Hollow Fiber Module

Conceptually, both capillary modules and hollow fiber modules are identical, differing only in their respective dimensions. In the hollow fiber module, the feed stream may flow either outside the hollow fiber or through the lumen of the fiber. When the module is operated with the feed flowing outside the hollow fiber and the permeate passing inside the lumen, the goal is to increase the overall membrane area and reduce the total pressure drop. Conversely, when operated in reverse configuration, as in the pervaporation process, where the feed flows through the lumen of the hollow fiber and the permeate flows outside the fiber, the aim is to prevent an increase in permeate pressure inside the fibers. Figure 1.6 depicts the schematic representation of a hollow fiber module. The maximum possible packing density for a hollow fiber module is around 30,000 m^2/m^3. The module is preferred for handling clean feed, such as in pervaporation and gas separation. It can also be used for the desalination process, provided the feed stream undergoes pretreatment.

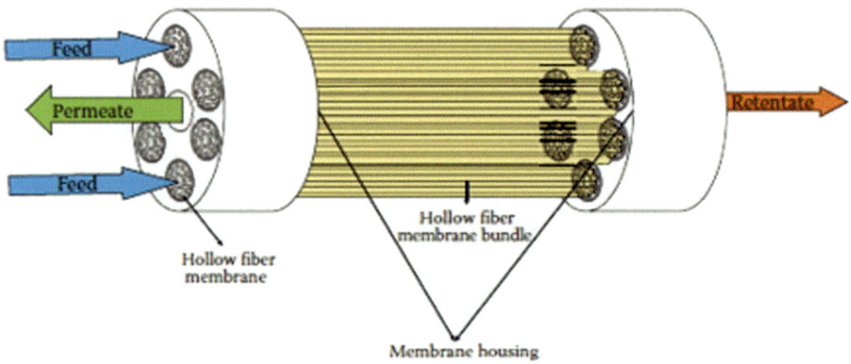

Fig. 1.6: Schematic representation of hollow fiber module

1.5 Mode of Operation

1.5.1 Batch Mode of Operation

Membrane based separation processes are usually carried out in the batch mode. In the batch mode of operation, a limited volume of feed is introduced to the membrane cell at a time, which is stored inside the cell prior to permeation. The permeation process will continue until the feed inside the cell is exhausted. Consequently, the feed solution needs to be refilled inside the membrane cell for further experimental runs. It operates identically to a batch reactor. This mode of operation is beneficial for membrane based separation processes, as the time for cleaning fouled membranes is available during the feed refilling step. Thus, it helps increase process efficiency and membrane shelf-life, while simultaneously reducing the expense and extent of membrane fouling. This mode of operation is widely applied in the pharmaceutical, biotechnology, and food industries. The schematic representation of a batch membrane based process is shown in Figure 1.7(a).

1.5.2 Semi-batch Operation

The semi-batch mode of membrane operation is almost identical to the batch mode but includes certain alterations to allow the addition of feed solution, removal of permeate/retentate or recycling of the permeate/retentate stream within the membrane setup. This helps to elevate both the selectivity and control of the membrane based separation process. Moreover, this mode of operation mode also reduces the load on downstream processing for both permeate and retentate, as the recirculation of permeate occurs multiple times during the membrane process. In process industries, semi-batch membrane processes are widely utilized. This operation mode also exhibits a lower propensity for membrane fouling and higher process efficiency. The schematic representation of a semi-batch membrane process is shown in Figure 1.7(b).

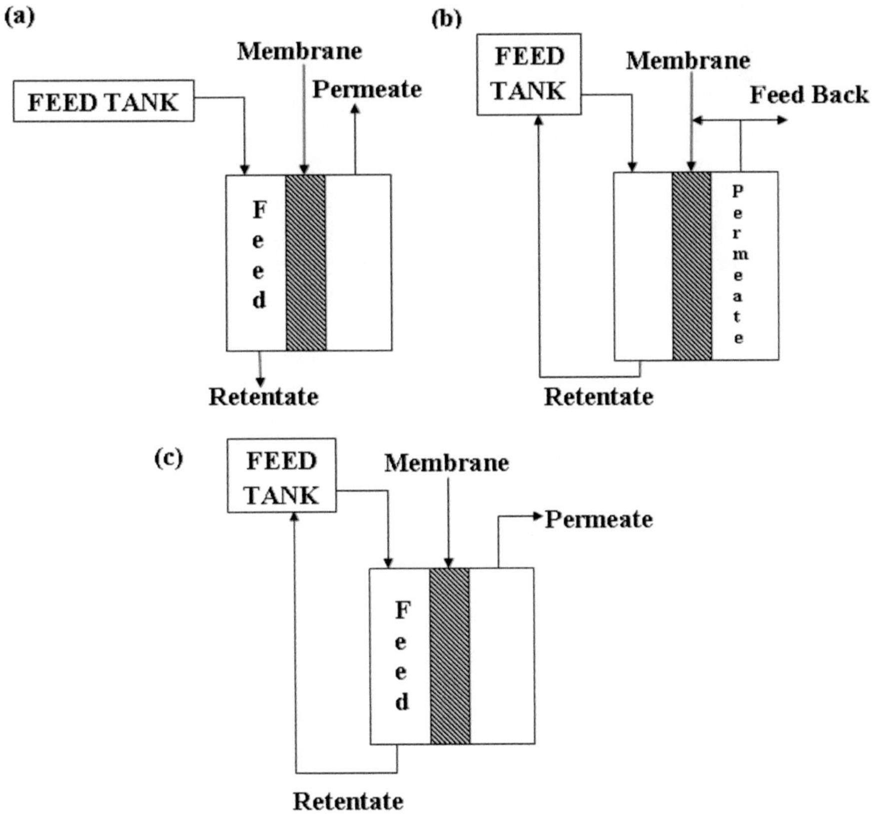

Fig. 1.7: Schematic representation of (a) batch; (b) semi-batch; and (c) continuous membrane based process

1.5.3. Continuous Mode of Operation

In continuous mode operation, both the addition of feed solution and the removal of permeate stream are carried out continuously. The capacity, speed, and efficiency of the membrane process in this mode of operation for handling a bulk volume of feed solution are usually high. Though this mode of operation mode lowers the operation time, it increases the probability of membrane fouling due to its continuous nature. This lowers the overall efficacy of the membrane separation process. The continuous mode of the membrane based process is schematically represented in Figure 1.7(c).

1.5.4 Other Modes of Operation

Depending on the flow direction of the feed and permeate streams, the two other common modes of membrane operation are dead-end and cross-flow filtration. These modes of membrane operation are widely adopted by various membrane-based processes across the globe. Dead-end filtration is most commonly utilized in batch mode membrane operation, whereas cross-flow filtration is preferred for the continuously operated membrane processes. Both have their inherent pros and cons. In the case of dead-end filtration, both the recovery and membrane fouling are higher. Due to fouling, the permeate flux gradually decreases over the course of operation time. On the other hand,higher recovery suggests better performance. However, the issue of fouling propensity is comparatively less in cross-flow filtration, and recovery efficiency is correspondingly lower. Controlling membrane fouling tendency suggests cross-flow as a choice for utilization. Thus, a hybrid membrane process enhanced with a synergistic amalgamation of both dead-end and crossflow filtration, can be promising, beneficial, and highly efficient in terms of operation and utilization. Both modes may be utilized either in single or multiple pass configurations, where the circulation of the feed stream or permeate across the membrane system occurs once or multiple times respectively.

1.5.4.1 Dead-End Filtration Mode

In dead-end filtration mode, the feed stream is introduced in a direction perpendicular to the membrane within the membrane system, as depicted in Figure 1.8(a). This mode of membrane operation is preferred for batch operation. Compared to cross-flow mode, its efficiency is higher. The major drawback of this mode is fouling, which reduces the efficiency of the process. Consequently, the membrane demands frequent cleaning during operation, which incurs additional expense and reduces the shelf-life of the membrane. As a result, these processes become somewhat expensive and less suitable for longer durations of use. Henceforth, the cross-flow mode of operation is usually preferred for longer-duration membrane-based processes.

1.5.4.2 Cross-Flow Mode

In case of cross-flow mode, the feed stream is introduced parallel to the membrane in any membrane system, as depicted in Figure 1.8(b). The popularity of cross-flow operation mode is due to its lower membrane fouling propensity. It helps reduce concentration polarization and fouling propensity to a significant level. Proper design of the membrane module, along with optimized cross-flow feed velocity can provide promising outcomes. These optimizations will facilitate higher mass transport and reduced fouling intensity in a membrane-based process. There are distinct configurations including counter-current, co-current, cross-flow, and perfect mixing. Amongst these, counter-current is the best configuration in terms of performance followed by the cross-flow, co-current flow, and perfect mixing, respectively.

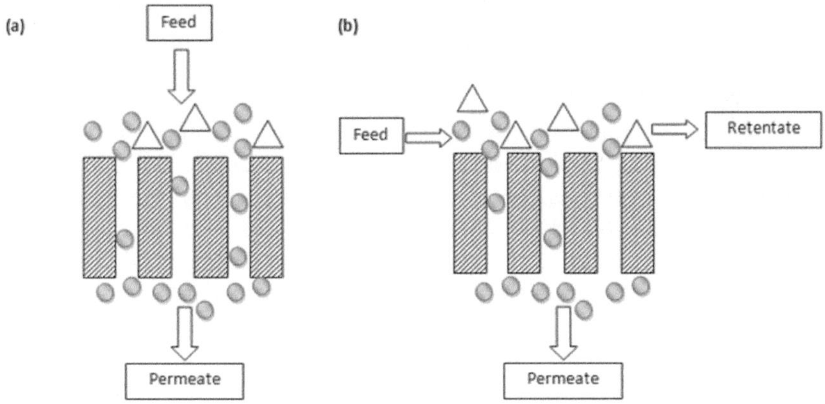

Fig. 1.8: Schematic diagram of (a) dead-end; (b) cross-flow membrane based separation process

1.6 Membrane Contactors

Membrane contactors are a type of membrane system where the membrane do not regulate the permeation rate of feed stream components but acts only as an interface in between the feed stream and permeate phase. The use of membrane contactors is mostly found in shell and tube membrane modules fitted with hollow fiber (microporous) membranes. Depending on the involved phases, various types of membrane contactors include gas–liquid and liquid–liquid membrane contactors. In the first category, feed exists in gaseous phase and permeate exists in liquid phase. However, in the second category, both the feed and permeate phases are in the liquid state, involving immiscible liquids. If the liquid-liquid process involves miscible liquids, then the process is called membrane distillation. These above-mentioned types of membrane contactors are represented schematically in Figure 1.9. These membrane contactors exhibit better performance than traditional liquid–liquid extractor or liquid–gas absorber as they offer the prime benefit of a very high contact area between two different phases. With around 10 times more contact area compared to a traditional tower of the same size, one of the best examples of this technology is the membrane blood oxygenator. It enables open-heart surgeries by bringing the essential amount of blood to an appropriate level. Another major benefit of membrane contactors is that no direct contact occurs between the two phases, as they remain separated by an interface. This reduces the chances of flooding or channeling during high-flow situations. Moreover, counterflow operation can be carried out to achieve optimal efficiency in separating or concentrating feed components or products. Thus, with a small quantity of extractant, the extraction of components from a bulk volume of feed is possible, resulting in higher effectiveness and economic feasibility of the process. Unlike traditional contactors, liquids with similar densities may be utilized, as the two phases do not come into direct contact. The primary disadvantages of membrane contactors stem from the membranes themselves. The transport rate is largely controlled by membrane properties, generally leading to slower separation rates. Membrane fouling is another significant drawback that further reduces the transport rate. Lastly, in some cases, the utilized membranes are unable to sustain elevated pressures, temperatures, and severe chemical conditions. Consequently, the utilization of membrane contactors becomes limited in various industrial separation processes. Thus, is a high demand to design and develop membranes with improved resistance to pressure, temperature, and harsh chemical environments.

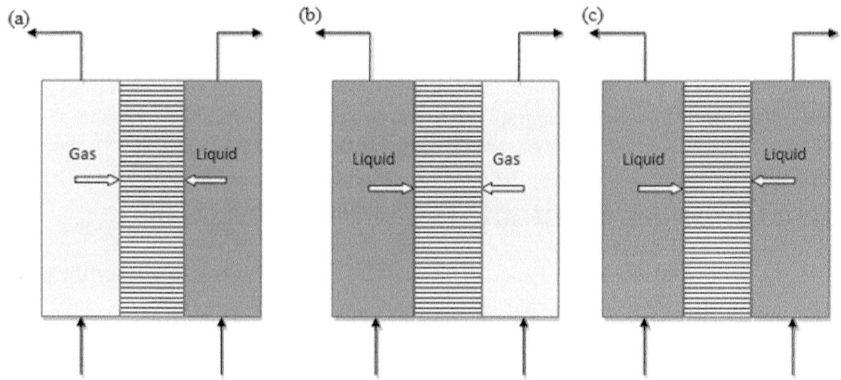

Fig. 1.9: Schematic representation of (a) gas-liquid; (b) liquid-gas; and (c) liquid-liquid membrane contactors

1.6.1 Types of Membrane Contactors

1.6.1.1 Gas–Liquid Membrane Contactors

Gas–liquid membrane contactors are the most commonly utilized membrane contactors for separating or transporting gases from the liquid phase. This type of membrane contactor is designed so that the gas phase and the liquid phase remain present on opposite sides of the membrane. Due to the existence of a partial pressure gradient, diffusion of the feed components occurs from the feed compartment to the permeate side or phase. One of the best examples of such gas–liquid membrane contactors is the blood oxygenator, which finds wide applications throughout the world. Pure air or oxygen (gaseous phase) flows on one membrane side whereas the blood (liquid phase) flows on the other membrane side. Pure oxygen diffuses into the blood stream from the gaseous phase, while carbon dioxide flows from the blood (liquid phase) to the gaseous phase, due to the existence of a partial pressure gradient in between the two different species. Both porous as well as non-porous dense membrane can be utilized for this type of membrane contactor. The membrane primarily acts as a barrier. The nature of the membranes, in terms of hydrophilicity and hydrophobicity, will determine whether the membrane pores will be wetted by a liquid or a gas. In hydrophilic membranes, the liquid phase wets the membrane pores. On the contrary, the gas phase wets the membrane pores in case of hydrophobic membranes.

1.6.1.2 Liquid–Liquid Membrane Contactors

Liquid–liquid membrane contactors use two different liquids: one as the feed stream and the other as the permeate stream, each on two opposite sides of the membrane. The components from the feed side are transported to the permeate side liquid through the membrane barrier. In case of liquid–liquid membrane contactors, the wetting or non-wetting behavior of the membrane pores depends on the types of liquids selected for the feed phase and permeate phase. For instance, if two chosen liquids are an organic solvent and an aqueous solvent, and the membrane is hydrophobic, then the membrane pores will be filled by the organic solvent. However, with a hydrophilic membrane, the opposite behavior will be observed. Instead of the organic solvent, the aqueous solvent will fill the membrane pores. These conditions allow the two distinct phases to maintain their individual profiles without any risk of intermixing due to their contrasting attributes. The interface for the transportation will remain on the side of the membrane with the non-wetting liquid. Notable applications of liquid–liquid membrane contactors include the removal and separation of hazardous pollutants, pesticides, volatile organic compounds, and bioproducts.

1.7 Membrane Reactor

A membrane reactor is essentially a hybrid combination of a reactor and a membrane, where the membrane is used for product separation, reactant distribution, or catalyst support. This combination allows for improved control, higher efficiency, and cost-effectiveness of a process. Depending on the application, various types of membrane reactors, equipped with different membranes, are utilized and operated in different modes. These membrane reactors are widely applied in various biotechnology, pharmaceutical, and chemical industries. Notable applications include simultaneous production and separation of valuable products in biotechnology and pharmaceutical industries, reaction and separation in the chemical industries, and purification of feed streams for environmental safety and security.

1.7.1 Principle of Operation

In the case of membrane reactors, both production and separation occur simultaneously, unlike traditional production processes where unit operations occur separately. Here, the synthesized products are separated during the process itself, reducing the load on the downstream processing steps. From now on, a membrane reactor is recognized as a hybrid setup that performs the synergistic functions of both a reactor and a membrane.In this arrangement, the membrane serves as reaction interface or separator, while the reactor maintains the essential operational conditions. This symbiosis between the reactor and membrane leads to a simple, economical, efficient, and safe process.

Depending on the application, membrane reactors utilize different types of membranes. Polymeric organic membranes are preferred for biochemical processes due to their inability to withstand elevated temperatures and harsh chemical environments. Consequently, these polymeric membranes are used in the production of products utilizing immobilized enzymes, the generation of biogas from waste, and ensuring environmental safety by treating polluted streams. Conversely, due to their ability to withstand higher temperatures and harsh chemical environments, inorganic membranes are preferred for catalytic membrane reactors. Despite their substantial potential for a wide variety of applications, membrane reactors face limitations related to the membranes themselves. Therefore, selecting membranes with the necessary properties of separation, toughness, compatibility, selectivity, and efficiency is a challenging task.

Membranes are employed in various ways within a membrane reactor to enhance operational effectiveness. Typically, in a membrane reactor, membranes can be utilized as product extractors, active contactors, or reactant distributors.

- *Product extractor:* For thermodynamically equilibrium reactions, removing a single product will shift the reaction equilibrium is removed, the equilibrium of the to the product side. Consequently, in membrane reactors for these types of reactions, the membrane acts as a product extractor and will thereby enhancing the efficiency of product conversion.Additionally, in some cases, the membrane functions as an extractor for an intermediate reaction product, preventing the formation of secondary or unwanted by-products. This enhances both the efficiency and selectivity of the process.
- *Active contactor:* The contact between the reactants and catalyst occurs in a controlled and uniform manner through the membrane. The diffusion of reactants across the membrane happens either from one side or simultaneously

from both the sides. This allows the reactants to access a higher number of active catalyst sites, thereby enhancing the catalyst's efficiency and favoring the process.

- *Reactant distributor:* The reactant necessary for a specific reaction or process is introduced at a very controlled rate using membrane reactors. On the feed side of the membrane, the reactant is gradually introduced to the reaction zone in a precisely controlled way. By restricting side reactions and preventing the formation of secondary by-products, this approach enhances both the efficiency and selectivity of the process. Additionally, impure reactants can be used, making the process more economical.

1.8 Summary

- Elaboration on the classification of membrane separation processes.
- Elaboration on the transport mechanism through the membrane discussing the solution-diffusion mechanism.
- Discussion on different transport models on membrane separation.
- Discussion on different membrane modules and the mode of operations.
- Elaboration on membrane contactors and reactors.

CHAPTER

2

Membrane Distillation

2.1 Introduction

In the previous chapter, we have discussed the applications of conventional membrane separation systems such as MF, UF, NF, and RO. These systems have widespread applications encompassing desalination and the food industry to the pharmaceutical industry, and ultimately, wastewater treatment. However, as we've observed, individual processes are limited by their inherent disadvantages. These limitations may arise from the costs associated with primary processes before the feed is introduced to a membrane unit or the costs related to the membrane itself. Additionally, membrane fouling, which depends on the effectiveness of the primary processes and the feed introduced to the membrane unit, is a major issue that further increases the cost of the membrane. Membrane distillation, a promising new technology envisaged in the 1960s, combines distillation with membrane separation to produce potable water from unpurified aqueous mixture. Distillation has been a popular process since ancient times. Aristotle (B.C. 384–322) stated "*Sea water can be made drinkable by vaporization; other liquids behave in the same way.*" The inclusion of a membrane leads to the concept of membrane distillation, which will be elaborated here.

2.2 Classification of Membrane Distillation Technology

A fundamental principle of any distillation process is the separation of two or more miscible liquid components from their mixture based on differences in their boiling point. However, a key obstacle in conventional distillation is the close boiling points of some liquids (forming azeotropes), which restricts their separation. In contrast, membrane distillation combines membrane separation technology with conventional distillation to avoid the complications arising from azeotrope formation. The membrane serves as a barrier, allowing only vapor to pass through based on its size. One of the critically acclaimed benefits of membrane distillation over conventional distillation is its energy efficiency, which will be elaborated on in the later section of this chapter through an energy analysis. Membrane distillation processes can be classified based on their basic operational techniques, such as direct contact membrane distillation (DCMD), air gap membrane distillation (AGMD), sweep gas membrane distillation (SGMD) and vacuum membrane distillation (VMD).

2.2.1 Direct Contact Membrane Distillation (DCMD)

In this process, the membrane acts as a contactor between the hot feed stream and cold water (Figure 2.1). Due to the temperature gradient, hot vapor passes through the hydrophobic membrane and comes in contact with the cold permeate stream, where it condenses. However, one of the primary disadvantages of DCMD is its its low thermal efficiency (high energy consumption per unit volume of distillate). Equation 2.1 shows the estimation of specific thermal energy consumption (STEC) for a membrane distillation process.

$$\text{STEC} = \frac{F_f \rho_f C_{p,f} \left(T_{f,inlet} - T_{f,outlet}\right)}{10^9 \times F_d \times t_{dc}} \left[\frac{MJ}{L}\right] \quad (2.1)$$

The primary mechanism in the membrane distillation process is the difference in partial pressure, with both mass and heat fluxes playing a crucial role in transporting the mass across the membrane. STEC is an important parameter in selecting the appropriate membrane distillation mode. In the DCMD process,obtaining a high amount of distillate decreases the STEC. As we increase the feed temperature, we approach the thermodynamic vapour-liquid equilibrium condition or the bubble point, which vaporizes the volatile product we want to recover as the distillate. Therefore, when the energy is supplied from external sources, it increases the distillate collection, ultimately reducing the STEC according to equation 2.1. However, increased feed temperature with compromised STEC applies to any distillation process, not just DCMD. Another mathematical parameter, called energy efficiency (EE), is used to assess the quality of the process, as given by equation 2.2.

$$\text{EE} = \frac{F_p \times \rho_p \times H_{lv}}{F_f \rho_f C_{p,f} \left(T_{f,inlet} - T_{f,outlet}\right)} \quad (2.2)$$

As the feed inlet temperature increases, efficiency also increases along with the distillate. This is the opposite of what is observed with STEC. While STEC reflects the energy consumption during the process, efficiency indicates the condensation efficacy. Therefore, a process may fail to meet the energy consumption criterion during membrane distillation, even if it operates with high efficiency.

2.2.2 Air Gap Membrane Distillation (AGMD)

In AGMD, unlike DCMD, the distillate passes through an air filled gap before it comes into contact with the condenser plate. Figure 2.2 shows a schematic for

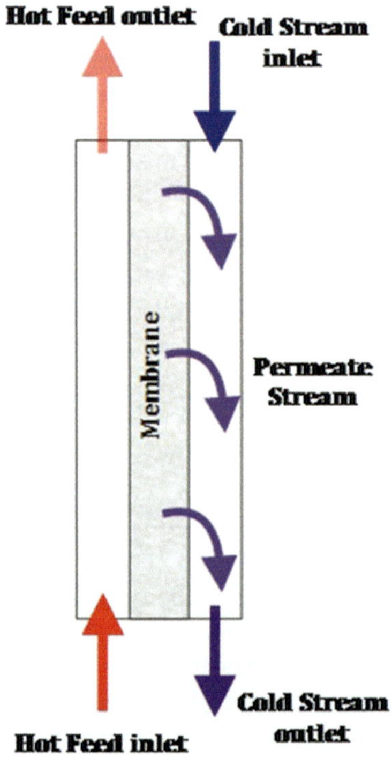

Fig. 2.1: Schematic for DCMD

the AGMD. One of the main advantages of AGMD over DCMD is the ability to control distillate collection after condensation. In DCMD, the condensation temperature depends on the final mixing temperature of the distillate and cold stream on the permeate side. If the mixing temperature is higher than expected, it reduces the vapour pressure difference between the feed and permeate sides. This results in a decreased pressure gradient across the membrane, leading to lower distillate production. Consequently, the EE decreases while the STEC increases. Conversely, in AGMD, condensation is influenced by the set temperature of the condensation plate,reducing reliance on the mixing temperature.

However, the main complexity with the AGMD lies in the content of the gap between the membrane contactor and the condenser plate. If this gap is filled with air, conductive loss is significantly reduced, but distillate collection becomes lower, leading to decreased EE. A critical issue in any membrane separation process is membrane fouling. AGMD shows lower fouling compared to DCMD. In membrane distillation, there are three main types of fouling – organic fouling, biofouling and inorganic fouling. During conventional size-based membrane separation processes, concentration polarization occurs due to the deposition of

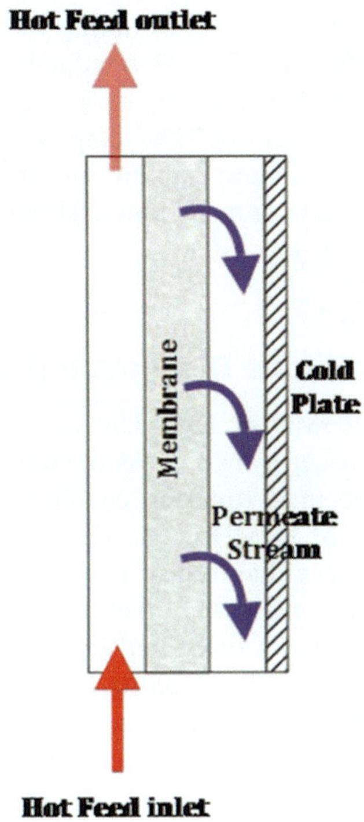

Fig. 2.2: Schematic for AGMD

solutes on the membrane, which contributes to fouling. During separation, solute particles accumulate on the membrane, creating a concentration gradient from the membrane surface to the bulk fluid. Such a negative gradient (equation 2.3) restricts the flow of the solute towards the surface of the membrane. Equation 2.3 shows the steady state mass balance equation across the membrane.

$$V_s C_p^* = V_s C_b^* - \left(-D_s \frac{dC^*}{dz}\right) A_m \tag{2.3}$$

While, with the membrane distillation along with the resistance an additional resistance forms due to temperature polarization that affects the separation process. Now if one looks at the mechanism for membrane distillation, for the transportation of the mass through the membrane, the bulk fluid temperature must be high enough in order to increase the vapour pressure during transportation. At the liquid-membrane interface on the permeate side, vapor condenses and part of the energy is released. Some of this energy is stored as latent heat for water vapour production, while the remainder is lost through the membrane material. This energy loss reduces the feed side temperature, subsequently lowering the partial pressure on the feed side. Conversely, as the energy is released on the permeate side, vapor pressure increases, resulting in higher vapor pressure on the permeate side toward the feed side. This reduction in the driving force for membrane distillation leads to decreased separation efficiency. This phenomenon is known as "Temperature polarization." The temperature polarization coefficient (TPC) is given by equation 2.4. Figure 2.3 illustrates the resistance effects caused by these two polarization effects in membrane distillation.

$$\text{TPC} = \frac{T_{fm} - T_{pm}}{T_{fb} - T_{pb}} \tag{2.4}$$

2.2.3 Sweep Gas Membrane Distillation (SGMD)

Figure 2.4 illustrates the SGMD process. Unlike AGMD, SGMD employs a moving cold gas stream to sweep away the volatile liquid that forms as permeate. This process enhances the mass transfer coefficient across the membrane. However, one disadvantage of SGMD is the increase in the temperature of the sweep gas along the membrane due to heat release at the gas-vapour interface on the permeate side. As the temperature rises, the vapour pressure difference is reduced, leading to a decrease in flux. Conversely, in SGMD, the moving gas stream ensures a constant driving force acting based on the partial pressure of the component. This consistently maintains a proper permeate flux across the membrane. Several experiments have shown that as the gas flow rate increases,

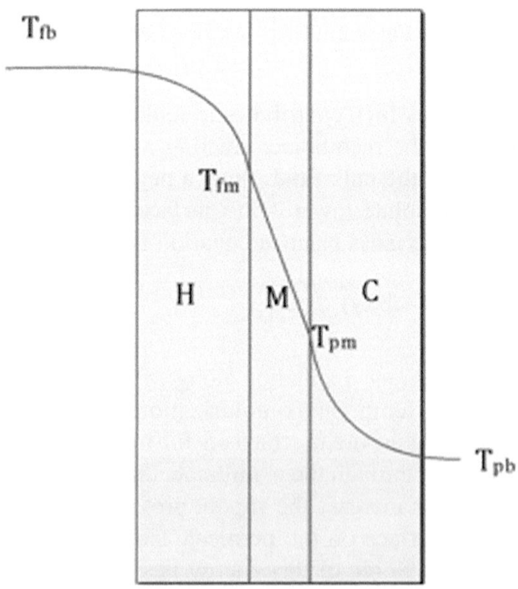

Fig. 2.3: Schematic for temperature profile in membrane distillation process

the permeate flux also rises due to an increase in the Reynold's number, which in turn enhances the heat transfer coefficient. However, the increase in permeate flux becomes asymptotic nature with the rise in gas flow rate. As the gas flow rate increases, the gas pressure on the permeate side also rises, creating a resistance to the permeate flow. Once the asymptotic nature is reached, any slight increase in gas flow rate will lead to a decrease in permeate flow. As the permeate mass transfer is the rate limiting step in SGMD, there should be an optimum gas flow rate to achieve moderate permeate flux.

In SGMD, heat transfer across the membrane occurs through three possible mechanisms.

(a) Heat transfer from the bulk liquid solution to the feed side hydrodynamic boundary layer adjacent to the membrane surface.
(b) Heat transfer through conduction across the membrane pores filled with gas molecules and through latent heat associated with the vaporized molecules across the membrane.
(c) Heat transfer through the permeate boundary layer.

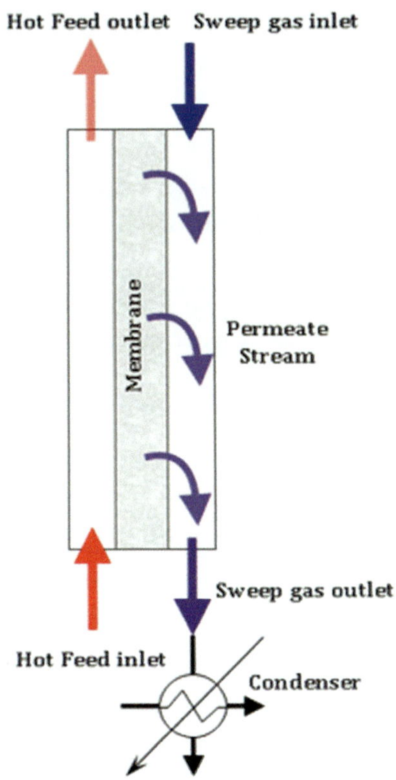

Fig. 2.4: Schematic for SGMD

$$Q_f = h_f \left(T_{fb} - T_{fm}\right) \quad (2.5)$$

$$Q_m = \frac{k_m}{\delta}\left(T_{fm} - T_{pm}\right) + \sum_{i=1}^{n} J_i \Delta H_{Vi} \quad (2.6)$$

$$Q_p = h_p \left(T_{pm} - T_{pb}\right) \quad (2.7)$$

At steady state condition (with no heat loss at the permeate side membrane surface), $Q_f = Q_m = Q_p$.

Therefore,

$$h_f \left(T_{fb} - T_{fm}\right) = \frac{k_m}{\delta}\left(T_{fm} - T_{pm}\right) + \sum_{i=1}^{n} J_i \Delta H_{Vi} = h_p \left(T_{pm} - T_{pb}\right) = K \quad (2.8)$$

Further on rearranging the equation 2.8

$$T_{fb} - T_{pb} = K\left(\frac{1}{h_f} + \frac{\delta}{k_m} + \frac{1}{h_p}\right) \text{ and } T_{fm} - T_{pm} = K\frac{\delta}{k_m} \quad (2.9) \text{ and } (2.10)$$

Now TPC for SGMD process can be given by,

$$\text{TPC} = \frac{\dfrac{\delta}{k_m}}{\left(\dfrac{1}{h_f} + \dfrac{\delta}{k_m} + \dfrac{1}{h_p}\right)} \quad (2.11)$$

Now as defined previously, EE is one of the important parameters that are required to assess the quality of the SGMD process. In this case, the net heat loss can be expressed as $\frac{k_m}{\delta}\left(T_{fm} - T_{pm}\right) + \sum_{i=1}^{n} J_i \Delta H_{Vi}$ against which the amount of latent heat transported with the vapour molecule towards the permeate side manifests as $\sum_{i=1}^{n} J_i \Delta H_{Vi}$.

Therefore, $$\text{EE} = \frac{\sum_{i=1}^{n} J_i \Delta H_{Vi}}{\frac{k_m}{\delta}\left(T_{fm} - T_{pm}\right) + \sum_{i=1}^{n} J_i \Delta H_{Vi}} \times 100 = \frac{\sum_{i=1}^{n} J_i \Delta H_{Vi}}{h_p\left(T_{pm} - T_{pb}\right)} \times 100 \quad (2.12)$$

As the gas flow rate increases, the heat transfer coefficient (h_p) on the permeate side also increases due its proportional relationship with Reynold's number. This results in a continuous increase in flux because of the rising thermal gradient. The mass transfer coefficient for a porous membrane can be controlled in two ways: one where it is governed purely by Knudsen diffusion and, and another where it is controlled by molecular diffusion. Thus the overall mass transfer coefficient is given by,

$$k_m = \frac{1}{\left(\frac{1}{k_m^{Kn}} + \frac{1}{k_m^{D}}\right)} \tag{2.13}$$

As air pressure increases, the molecular mass transfer coefficient decreases which is expected with an increase in air flow rate. Therefore, the overall mass transfer coefficient decreases, leading to an increase in the permeate flux but a reduction in EE. Qualitatively, EE is defined as the amount of separation achieved relative to the amount of heat loss. In this scenario, an increase in k_m leads to a corresponding rise in heat loss, which ultimately reduces the EE.

2.2.4 Vacuum Membrane Distillation (VMD)

In VMD, the permeate side is maintained at a pressure lower than the equilibrium vapour pressure. One typical advantage of VMD is low heat conduction through the membrane, owing to the low vapor pressure on the permeate side. Consequently, the primary driving force for membrane transport in VMD is the pressure difference, leading to Knudsen transport and Poiseuille flow. However, in a typical VMD process, the radii of membrane pores are well below the mean free path of the permeating molecule. Therefore, in the VMD process, Knudsen transport is the dominant transport mechanism and the molar flux is described by equation 2.14.

$$J_i = \frac{K_m}{\sqrt{M_i}}\left(P_{fm} y_{i,fm} - P_{pm} y_{i,pm}\right) \tag{2.14}$$

The total flux is given by equation 2.15 after summing up all the components.

$$J = \frac{K_m}{\sqrt{M}}\left(P_{fm} - P_{pm}\right) \tag{2.15}$$

where average molecular weight $\sqrt{M} = \frac{\sum_i J_i \sqrt{M_i}}{J}$ (2.16)

Energy transport across the membrane is given by equation 2.17 after combining 2.5 and 2.6.

$$Q_f = h_f\left(T_{fb} - T_{fm}\right) = \frac{k_m}{\delta}\left(T_{fm} - T_{pm}\right) + \sum_{i=1}^{n} J_i \Delta H_{Vi} \tag{2.17}$$

Adopting an approach similar to that taken for equation 2.8,

$$T_{fb} - T_{pm} = \left(\frac{1}{h_f} + \frac{\delta}{k_m}\right) \text{(Considering insignificant latent heat of vaporization)} \tag{2.18}$$

In VMD, the sensible energy loss on the permeate side is negligible, reducing the loss below 3% of the total energy loss. Therefore, sensible energy loss is typically disregarded in the VMD process.

2.3 Wetting of the Membrane

One of the prime concerns in the membrane distillation process is membrane wetting. As discussed earlier, the vapor must penetrate the membrane pores for effective separation in membrane distillation. Therefore, preventing membrane wetting is crucial for achieving prominent separation. Quantification wetting is essential for assessing the efficiency of membrane distillation. Liquid entry pressure (LEP) is the key parameter used to evaluate performance. LEP, of paramount interest in the membrane distillation process, is defined as the force required to allow water to penetrate the structure and is given by the Young-Laplace type equation.

$$\Delta P = -\frac{2\gamma \cos\theta}{r} \tag{2.19}$$

However, equation 2.19 is valid for axially regular pores, while there is alteration with the axially irregular pores because of the structural angle as shown in the Figure 2.5. In this case, equation 2.19 can be remodified with equation 2.20.

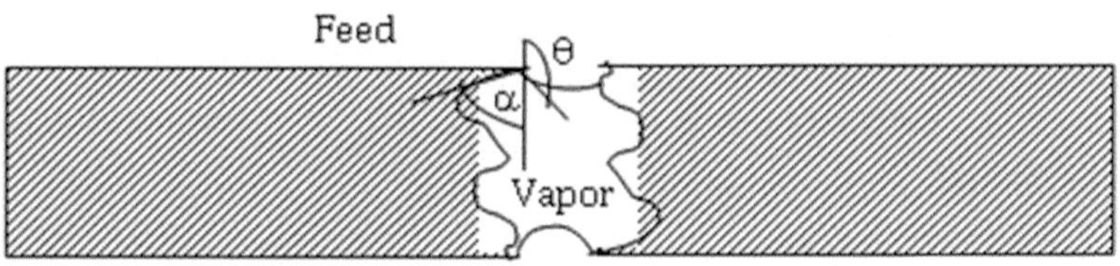

Fig. 2.5: Angular component arising in with the axially irregular pore

$$\Delta P = -\frac{2\gamma \cos(\theta - \alpha)}{1 + \frac{r}{\gamma}(1 - \cos\alpha)} \tag{2.20}$$

One of the primary considerations in membrane distillation is to ensure an increase in the contact angle, which eventually reduces LEP and subsequently enhances process efficiency. Another primary concern is the surface tension between the feed solvent and the membrane material. According to thermodynamics, the presence of organic contaminants decreases the surface tension compared to pure water, which is advantageous for the membrane distillation process as it enhances the separation of organics from its solution. Selecting the appropriate membrane material is crucial for effective separation using membrane distillation. Additionally, the LEP value of the membrane is influenced by the feed solution used in the membrane distillation process. For instance, consider dye removal from textile wastewater using membrane distillation. Silva et al. (8) conducted a detailed study on the observation of LEP with textile dye solution. They used polytetrafluoroethylene (PTFE), polypropylene (PP) and polyvinyl difluoride

(PVDF) membrane (of pore size around 0.2 μm) for the separation. They prepared a synthetic solution containing various dyes, including reactive black (Tiafix RBL 20505), disperse black (Trillon KCR), acid black (Trimacid CP 194), and direct black (Tricel NG-LBR 22). Table 2.1 shows the LEP observed with different types of dyes in comparison to water. The table indicates that with the dye solution, LEP decreases compared to water, creating more favorable conditions for membrane distillation. Therefore, when selecting a membrane for this process, it is effective to compare the LEP quoted by the manufacturer with the LEP of pure water. Table 2.2 presents the LEP values for various commercially available membranes, clearly indicating that if LEP is the primary criterion for membrane selection in membrane distillation, TF1000 would be an excellent choice.

Table 2.1: LEP (bar) observed with different types of dyes

Solution	PTFE	PP	PVDF
Water	3.6	1.2	3.1
Reactive black	2.3	1.2	2.1
Disperse Black	2.8	1.2	2.4
Acid Black	0.2	0.4	0.2
Direct Black	2.0	0.4	1.2

Table 2.2: LEP (bar) for different market available membranes (9)

Trade name	Manufacturer	Membrane material	Support material	Mean pore size (μm)	LEP (bar)
TF200	Gelman	PTFE	PP	0.20	2.82
TF450	Gelman	PTFE	PP	0.45	1.38
TF1000	Gelman	PTFE	PP	1.00	0.48
Emflon	Pall	PTFE	PET#	0.02	15.85
Emflon	Pall	PTFE	PET	0.20	5.51
Emflon	Pall	PTFE	PET	0.45	2.06
FGLP	Millipore	PTFE	PE*	0.20	2.80
FHLP	Millipore	PTFE	PE	0.50	1.24
GVHP	Millipore	PVDF	None	0.22	2.04
HVHP	Millipore	PVDF	None	0.45	1.05

#Polyethyleneterephthalate (PET); *Polyethylene (PE)

2.4 Applications of Membrane Distillation

Nowadays, membrane distillation has widespread applications. Wastewater treatment is one of the most promising applications of membrane distillation, which operates as a non-isothermal process to separate water from pollutants. Currently, the most commonly used membrane separation techniques for producing potable water from wastewater include reverse osmosis, following ultrafiltration and nanofiltration as described in the previous chapter. However, the primary drawback of these membrane processes is the osmotic pressure limitation, which reduces the permeate flux and makes the process less economical. On the contrary, thermally driven membrane processes like membrane distillation can reduce such limitations. The application of a specific membrane distillation process depends on the nature of the pollutants that that need to be removed from wastewater. For example, if the pollutants are primarily non-volatile solutes, DCMD is an appropriate separation method for water. Consequently, DCMD is employed for desalination to separate salt from water. Let's consider an example. Cooling tower blowdown water (CTBW) is a concentrated mixture of ions such as carbonate, bicarbonate, silica,and calcium. This concentrated stream cannot be recycled back to the cooling tower as makeup water because it forms scale over the heat transfer area. Therefore, the hardness of the stream must be removed to use it as makeup water. DCMD, using a hydrophobic membrane, is employed for this purpose. Let's correlate the limitations here with the use of a hydrophobic membrane. When a hydrophobic membrane is used, water permeation is reduced[1]. Therefore, the osmotic limitation that encountered in reverse osmosis can be managed and reduced. The DCMD process can remove almost 99% of salt with moderate flux. A typical limitation during the reduction of hardness through DCMD is the formation of scale on the membrane if insoluble carbonate is present in water.

One of the most common applications in wastewater treatment is the activated sludge process. In this process, wastewater is treated in a chamber with microorganisms that degrade biodegradable substances (Figure 2.6). With the inclusion of a membrane bioreactor, pretreated water from the chamber is fed to the membrane, where the microorganisms are separated and returned to the chamber. However, one of the critical issues with the process is the passage of low molecular weight components, whether biodegradable or non-biodegradable, which impacts the BOD and COD of the wastewater. Recently, there has been growing interest in the membrane distillation bioreactor (MDBR), which retains all types of non-volatile components such as salts, micro-organisms, and non-volatile organics. Figure 2.7 shows a schematic set-up for the MDBR, where the permeate stream from membrane (M1) is heated to a certain degree before being fed to the membrane distillation unit (M2). This eventually reduces the passage of non-volatiles, which passes through M1 because of its low molecular weight. 70-80% BOD removal can be achieved with the MDBR process. However, one of the critical features with the MDBR is the formation of biofilm that interacts with the inorganic ions and nutrients that lead to a continuous growth over the membrane surface. Fouling over MDBR is more pronounced due to the secreted extracellular polymeric substances from the microorganisms that interact with ions leading to crystallization followed by the formation of an impermeable crust on

[1]Here water permeation refers to its liquid state. However, vapor goes to the permeate side unchanged.

the membrane. This increases the mass transfer resistance through the membrane, leading to a reduction in flux. However, MDBR techniques can effectively treat pharmaceutical wastewater, municipal wastewater, and wastewater from the cosmetic industry.

Azeotropic distillation is one of the critical and costly processes within separation techniques. Readers may have questions, which can be addressed through a brief note on thermodynamics, relating it to the membrane separation process. Let's first address the basic questions that may arise.

- What is an azeotrope?
- What is distillation and why is it affected by azeotropes?

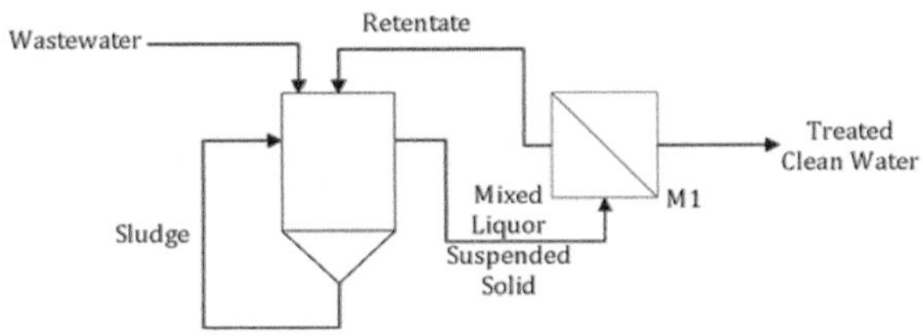

Fig. 2.6: Schematic for activated sludge process

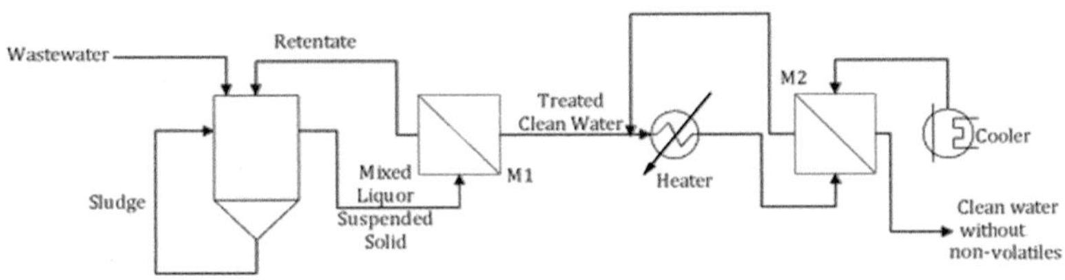

Fig. 2.7: Schematic of MDBR

Let's start with the second to better explain the first. Distillation is a process where miscible components in a liquid mixture are separated based on the boiling point differences between them. Let's take one example of a binary mixture of A and B, where the boiling point of A (solvent) is T_{BA} and the boiling point of B (solute) is T_{BB}. Now if $T_{BA} >> T_{BB}$, the B component will be separated as a top product of the distillation column and A will come out as the bottom product of the column (Figure 2.8). As B has a lower boiling point, it vaporizes quickly compared to A. However, when the boiling points of two components become almost the same after some separation time, no further separation can be achieved based on boiling point differences. This condition is known as an azeotrope. Consequently, the vapour pressure reduces for each component as B is separated. Eventually, the combined vapor pressure of A and B equals or exceeds the vapor pressure of the mixture. In other words the boiling point of the azeotrope is the same as the boiling point of the solution. Equation 2.19 represents Raoult's law which underpins the separation principle. In conventional distillation processes, the azeotropes cannot be separated without using a method called extractive distillation. In this process, one of the components (typically the non-volatile one) dissolves in a third solvent (miscible, high boiling, relatively non-volatile) (*S*), altering the property of *A*. Consequently, *B* appears in greater quantity as the top product, while *A* along with the third solvent, is separated as the bottom product. Figure 2.9 shows a schematic for the extractive process. However, the primary disadvantage with the extractive distillation process is the recovery of *A* from the third solvent. This needs a further downstream treatment that finally increases the process cost.

$$\text{Partial pressure of a component} = \text{Mole fraction} \times \text{Vapor pressure of pure compenent} \quad (2.21)$$

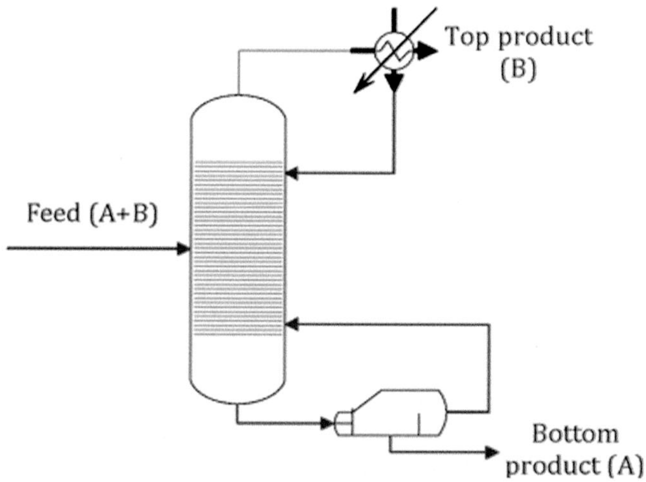

Fig. 2.8: Basic schematic for normal distillation process

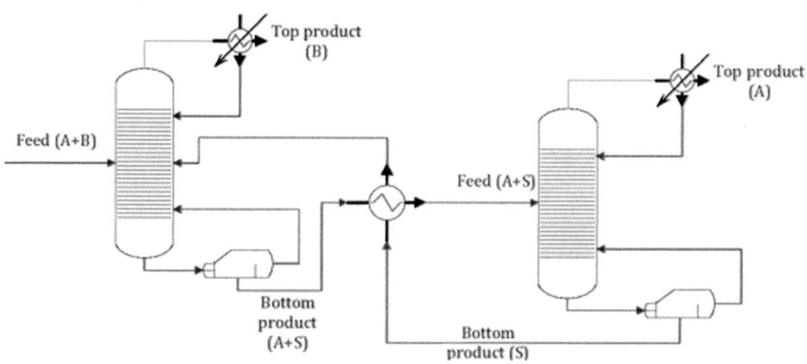

Fig. 2.9: Basic schematic for extractive distillation process

Membrane distillation could be an alternative method to avoid azeotrope formation. Before discussing membrane distillation for azeotropes, it is crucial to consider selectivity as an important parameter. Equations 2.20 to 2.23 illustrate the mass transport criteria across the membrane.

$$J_i = C_i^* \Delta p_i \tag{2.22}$$

$$\beta = \frac{y_i}{x_i} = \frac{J_i / M_i}{\sum J_j / M_j} = \frac{1}{x_i}\left[\frac{1}{1+\left(\frac{C_k^*}{C_i^*}\right)\left(\frac{M_i}{M_k}\right)\left(\frac{\Delta p_k}{\Delta p_i}\right)}\right] \tag{2.23}$$

In AGMD, the transport can be approximated with molecular diffusion. In this case, the molar transport across the membrane is proportional to the molecular diffusion.

$$C_i \alpha D_i M_i \tag{2.24}$$

$$\beta = \frac{1}{x_i}\left[\frac{1}{1+\left(\frac{D_k}{D_i}\right)\left(\frac{\Delta p_k}{\Delta p_i}\right)}\right] \tag{2.25}$$

The quantity β is independent of any thermodynamic relationship. Instead, it relies on the diffusivity of the substance, which in turn depends on the affinity and the partial pressure difference across the membrane. Therefore, separation after azeotrope formation can be adjusted, allowing separation even after azeotrope formation. This suggests that the β value can increase beyond 1. Figure 2.10 illustrates a basic advantage of shifting the azeotropic point, where β = 1 applies to Raoult's law.

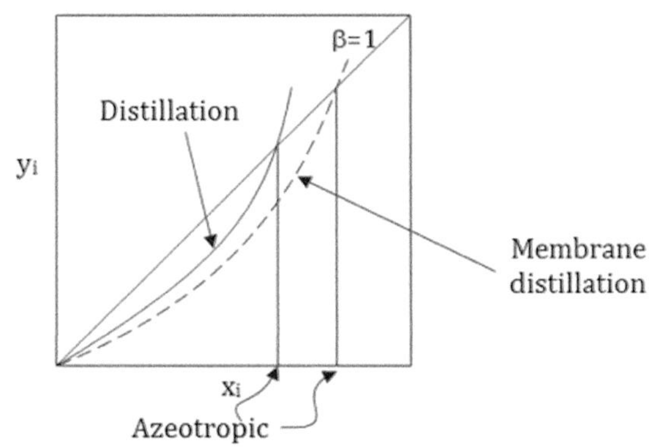

Fig. 2.10: Shifting of azeotropic point

2.5 Fabrication of Membrane for Membrane Distillation

A comprehensive discussion on the membrane distillation process and its applications has been provided. From the previous section, it is clear that the material of the membrane is specific to its application and designed accordingly. Various membranes are utilized in the membrane distillation process. Table 2.3 presents some membrane applications along with their specific materials for membrane distillation.

One of the widely recognized critical factors in the membrane distillation process is the hydrophobicity of the membrane, along with the specific membrane distillation configuration employed. Consequently, the most commonly used membranes in this process are PVDF, PTFE and polypropylene (PP) membranes. The performance of the membrane in the membrane distillation process largely depends on its preparation and the membrane material. In a study (13), dilute ammonia solution was introduced through a hollow fibre membrane, and SGMD was employed to separate ammonia. PVDF and ethylene glycol were mixed with two different solvents – N-methylpyrollidone (NMP) and N,N-dimethylformamide (DMF). The two solvents demonstrated distinct performances. However, preparation process parameters such as polymer extrusion rate and winding speed also influenced the ammonia flux. Ammonia removal was found to be highest with the membrane prepared by dissolving polymer in NMP solvent. Figure 2.11 illustrates the structure of NMP and DMF. In the case of NMP, the molecule arranges in parallel with anti-parallel polar group (14), resulting in an intra bond arrangement with increased hydration.

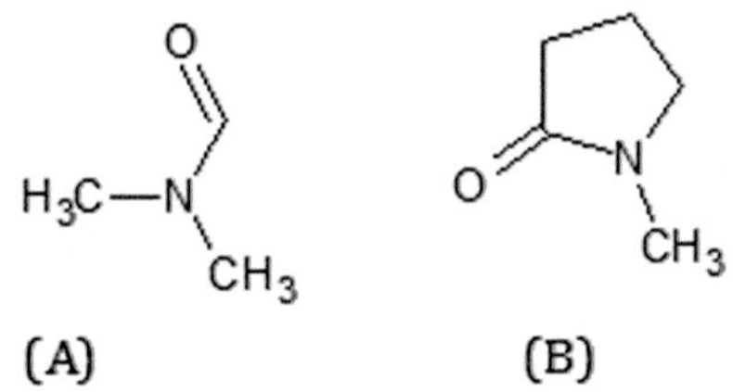

Fig. 2.11: (a) DMF solvent; (b) NMP solvent

Fabricating membranes for membrane distillation is a primary concern, as several factors must be considered to enhance the membrane's energy efficiency and throughput. Additionally, since membranes are always under thermal stress during distillation, their mechanical strength is also a critical issue. Additionally, aside from the previously mentioned membranes, PES membrane is another option that may be used for membrane distillation. The challenge lies in the hydrophilicity of the membrane. Increasing its hydrophobicity will enable the use of PES in membrane distillation, particularly with the DCMD configuration.

Table 2.3: Membrane distillation application specific membrane material

Membrane	**Membrane distillation configuration**	**Feed**	**Feed temperature (°C)**	**Reference**
Polytetrafluoroethylene (PTFE)	AGMD	HCl-Water (20.8:79.2 by weight)	50	(10)
Polyvinylidene fluoride (PVDF)	DCMD	Synthetic oilfield	65	(11)
Nanofibrous styrene-acrylonitrile (SAN)-hydrophilic non-woven	DCMD	Textile wastewater	60	
Negatively charged PVDF	DCMD	Pharmaceutical wastewater	60	
Anaerobic Membrane Bioreactor-Membrane Distillation	DCMD	Synthetic municipal wastewater	35	(12)

Nonsolvent-induced phase separation (NIPS), also known as "wet-phase inversion," is a commonly used technique for membrane fabrication. In this process, a homogeneous polymer solution is cast into a film and then immersed in a nonsolvent bath, where the exchange of nonsolvent and solvent between the bath and the film enriches the film in nonsolvent. A phase separation of the film occurs, resulting in a polymer-rich phase, known as the membrane matrix and a polymer-poor phase, which forms the membrane pores. However, the detailed mechanism of NIPS remains unclear, and fabricators often rely heavily on experience. During NIPS, the polymer used for membrane preparation is initially dissolved in an organic solvent to form a homogeneous mixture. The mixture is then cast into a film and immersed in a nonsolvent, which is a solvent in which the polymer is not miscible. Initial coagulation occurs rapidly on the top layer of the film, creating a barrier that restricts the percolation of the nonsolvent into the film and prevents solvent. As a result, macro voids form inside the membrane, while the top of the film has small pores. Figure 2.12 illustrates a schematic for the NIPS technique. After fabricating the PES membrane, it is grafted with a water repellent group to make it hydrophobic, allowing it to be used for membrane distillation.

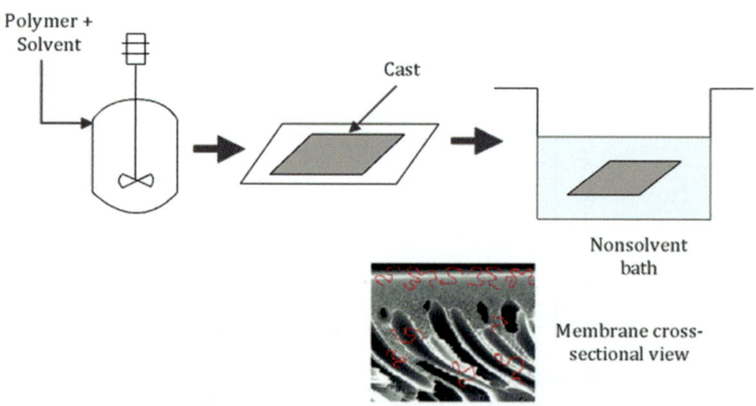

Fig. 2.12: NIPS process for the fabrication of the membrane

Another technology used for membrane preparation is thermally induced phase separation (TIPS). In TIPS, the polymer solution is dissolved in a solvent to create a homogeneous mixture at an elevated temperature. Unlike, NIPS, TIPS involves casting the membrane support by reducing the temperature. The criterion for selecting the solvent in TIPS is that the solvent, also called a "latent solvent", does not dissolve the polymer at room temperature. Rather, it dissolves the polymer at its melting point. To understand the TIPS process, one important parameter is the crystallization temperature, which is the temperature at which crystals start to appear during the cooling of a solution. During phase separation in TIPS, there are two possibilities. One is the appearance of polymers as crystals and the second, is the separation of the polymer in the presence of a nonsolvent with delayed crystallization. The first scenario represents the S-L (solid-liquid) separation while the second is L-L (liquid-liquid) separation, which is primarily used to fabricate the membrane film. Figure 2.13 presents a phase diagram for both cases.

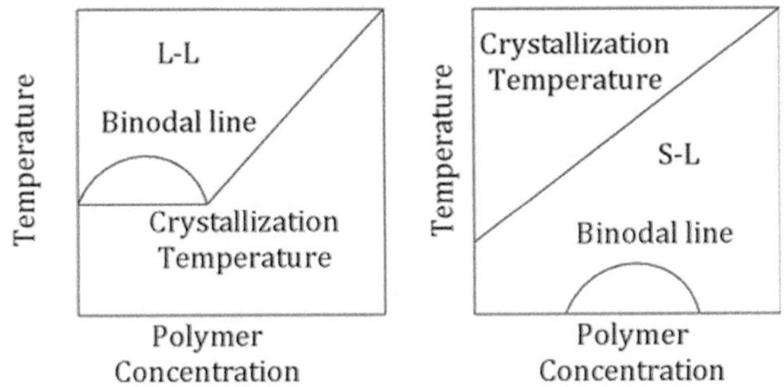

Fig. 2.13: Phase diagram during TIPS

During phase separation, a decrease in temperature leads to phase separation, illustrated by the binodal line in Figure 2.13. Without a nonsolvent, polymers form a homogeneous mixture with the solvent (diluent) and crystallize upon cooling (right side of Figure 2.13). With a nonsolvent, the polymer has low compatibility with the diluent and forms a film (polymer lean phase) instead of nucleating as crystals. At the crystallization temperature, tiny crystals begin to form. However, TIPS is often preferred for membrane distillation due to its advantages, such as s low defects, controlled pore sizes, and the formation of porous symmetric membranes. Although NIPS is commonly used because of its simplicity,. TIPS provides greater mechanical strength, making it advantageous for membrane distillation.

2.6 DCMD and Utilization of Solar Energy Towards Filtration of Brackish Water (15)

Today, one of the biggest global issues is the scarcity of potable water. Additionally, the increase in the global carbon footprint,raising Earth's temperature, poses a significant threat to humanity. Furthermore, using conventional energy resources for water filtration raises process costs, which are ultimately passed on to consumers. Using natural energy resources to address water scarcity not only solves the environmental problems but also tackles the issue of water scarcity. The integrated solar PV-DCMD process manifest a process involves heating brackish water with a PV cell and feeding it a membrane distillation for vapor permeation and collection (16). Figure 2.14 shows the schematic of the integrated PV-DCMD process for water purification. The primary driving force for membrane permeation is the temperature difference across the membrane. The feed side temperature depends on irradiance levels, directly influencing process performance. Equation 2.15 estimates the filtration rate per unit membrane area, and vapor pressure can be determined using the Antoine equation (equations 2.26 and 2.27).

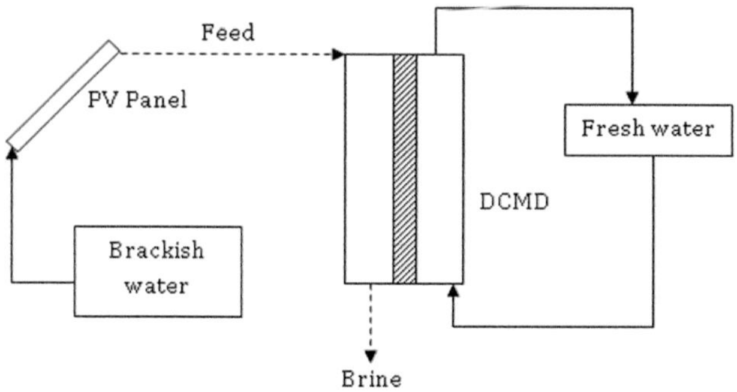

Fig. 2.14: PV-DCMD process for water purification

$$P_{fm} = \exp\left(23.328 - \frac{3841}{T_{fm} - 45}\right) \tag{2.26}$$

$$P_{pm} = \exp\left(23.328 - \frac{3841}{T_{pm} - 45}\right) \tag{2.27}$$

In DCMD, there will be slight modifications to equations 2.8-2.10 and 2.17-2.18. One must consider the latent heat of vaporization on the permeate side. In VMD, it's reasonable to assume the latent heat of vaporization is zero, as the condensation initiates the vacuum. In SGMD, the sweep gas carries away permeate to create a vapor pressure difference across the membrane. In direct contact membrane distillation, vapor pressure creation relies on the latent heat of vaporization on the permeate side. Efficient energy conduction from the feed side to permeate side reduces the vapor pressure difference across the membrane. This reduction in performance is assessed using equations 2.28 and 2.29, which evaluate the temperatures on the feed and permeate sides, considering the latent heat of vaporization on the permeate side.

$$T_{mf} = \frac{k_m\left(T_{pb} + \left(\frac{h_f}{h_p}\right)T_{fb}\right) + \left(\delta\left(h_f T_{fb} - J\Delta H_V\right)\right)}{k_m + \left(h_f\left(\delta + \frac{k_m}{h_p}\right)\right)} \tag{2.28}$$

$$T_{mp} = \frac{k_m\left(T_{fb} + \left(\frac{h_p}{h_f}\right)T_{pb}\right) + \left(\delta\left(h_p T_{pb} - J\Delta H_V\right)\right)}{k_m + \left(h_p\left(\delta + \frac{k_m}{h_f}\right)\right)} \tag{2.29}$$

Over the membrane surface within the channel, the flow is laminar. Therefore, the heat transfer coefficient can be calculated using the Sieder-Tate equation (equations 2.30 and 2.31) for the feed and permeate sides, respectively.

$$Nu_f = 0.027\left(Re_f\right)^{0.8}\left(Pr_f\right)^{0.33}\left(\frac{\mu_{fb}}{\mu_{mf}}\right)^{0.14} \tag{2.30}$$

$$Nu_p = 0.027\left(Re_p\right)^{0.8}\left(Pr_p\right)^{0.33}\left(\frac{\mu_{pb}}{\mu_{mp}}\right)^{0.14} \tag{2.31}$$

The energy transfer due to irradiance mainly creates a temperature difference between the bulk feed and the feed side membrane. This depends on factors such as the salt concentration in the brackish water. Key issues in heating the feed stream arise during the heat exchange operation. High salt concentrations can cause tube fouling, reducing the heat exchange efficiency. The overall heat transfer coefficient is given by equation 2.32.

$$\frac{1}{U} = \frac{1}{h_i} + R_{f_i} + \frac{\ln\left(\frac{D_o}{D_i}\right)}{2\pi K_H L} \tag{2.32}$$

The performance, quantified as evaporation efficiency (EE), is given by equation 2.33.

$$\text{EE} = \frac{J\Delta H_V}{U\left(T_{fb} - T_{mf}\right)} \times 100 \tag{2.33}$$

Increasing the salt concentration in brackish water can enhance the heat transfer coefficient. Liu et al. (17) observed a 50% increase in the heat transfer coefficient with an 8% increase in the salt concentration. As per equation 2.32, this raises the overall heat transfer coefficient and the feed-membrane contact temperature. Consequently, the feed temperature significantly impacts the process efficiency. It was noted that raising the feed temperature from 30°C to 45°C increases EE from around 35% to 42% (16). However, increasing the feed temperature can also lead to temperature polarization in VMD, which complicates separation. While TPC affects VMD significantly, DCMD experiences less impact. In DCMD, TPC affects both sides of the membrane due to vaporization occurring on both sides (18). On the contrary, TPC effect can be felt on the feed side interface in case with the VMD. Let us go with some of qualitative description on this especially in case with DCMD. A temperature decrease on the interface reduces vapor pressure on the feed side, while accumulated heat on the permeate side increases vapor pressure. In VMD this effect is linear. However, in DCMD, it depends on factors like flow direction and flow rate (19). In counter-current flow, where the feed channel inlet and permeate channel exit are close, mixing is improved on the feed side, reducing the likelihood of temperature polarization. While on the contrary, the permeate side exit temperature is higher than that of the inlet, but not more than that of the feed side. Hence, development of temperature polarization is not enunciated due to less heat passage from the feed side to the permeate side. However, with the concurrent system, the scenario attributes to higher possibility of temperature polarization that eventually restricts the permeation. While on the contrary, the permeate side exit temperature is higher than that of the inlet, but not more than that of the feed side. Hence, development of temperature polarization is not enunciated due to less heat passage from the feed side to the permeate side. In a concurrent system, there is a higher chance of temperature polarization, restricting permeation. Conversely, the permeate side exit temperature is higher than the inlet but lower than the feed side, resulting in minimal temperature polarization due to reduced heat transfer from the feed side. In a concurrent arrangement, immediate contact between the cold permeate and the hot feed results in rapid heat loss through the membrane. This reduces the partial pressure on the feed side and increases on the permeate side, hindering mass transfer.

Additionally, temperature polarization is influenced by the flow rate of the feed and permeate, as it affects the contact time through the membrane. Higher flow rates ensure efficient mixing reducing the likelihood of temperature polarization. Proper mixing thins the thermal boundary layer, reducing the temperature drop at the feed-membrane interface. However, less contact time between the permeate and feed through the membrane decreases energy transfer and partial pressure build up on the permeate-membrane interface (19). One primary advantage of DCMD is the simplicity in membrane module design and heat recovery. However, VMD is the most efficient among all membrane distillation processes (20). Solar-integrated membrane distillation process for potable water production is a promising technology currently being implemented. Numerous studies have been published on this. However, the main challenges are heat recovery and compensating for energy depletion due to low irradiance levels. Selecting and implementing the appropriate membrane distillation technology is crucial. Some processes offer high efficiency but poor recovery, and vice versa. Understanding the necessary trade-offs is essential for making the right choice. The salinity of water is a key factor in determining the cost of producing potable water from brackish water. In VMD, electrical and thermal energy costs are much lower compared to DCMD due to better thermal recovery. In DCMD, thermal energy costs increase, but higher salinity improves the convective heat transfer coefficient, enhancing energy flow across the membrane. The cold permeate in countercurrent DCMD creates a partial pressure difference for permeation. The process slows at the permeate exit, where temperature build-up enhances thermal recovery. Conversely, VMD uses vacuum-induced permeation with less thermal recovery but higher mass transfer and thermal efficiencies (20).

2.7 Summary

- An elaboration on different membrane distillation processes along with the governing equations relating mass and energy transfer across the membrane.
- Characteristics of the membrane used in membrane distillation, especially the wetting capacity of the membrane.
- Concept of MDBR.
- Shift of the azeotropic point in membrane distillation offers advantages over normal azeotropic distillation.
- Fabrication strategy for the membranes for membrane distillation.
- Application of DCMD in production of potable water from brackish water and different operating parameters that actually influence the efficiency of the process.
- Techno-economical assessment of DCMD in comparison to VMD.

CHAPTER

3

Electrodialysis

3.1 Introduction

In Chapter 1, we discussed RO membrane technology for salt ion removal. RO requires high pressure across the membrane to achieve moderate permeate flux. Implementing RO technology is costly. Electrodialysis is an alternative, using electrical driving force for ion separation. Electrical energy is a key cost, along with specialized membranes for specific ions. Figure 3.1 provides a preliminary schematic of the electrodialysis process.

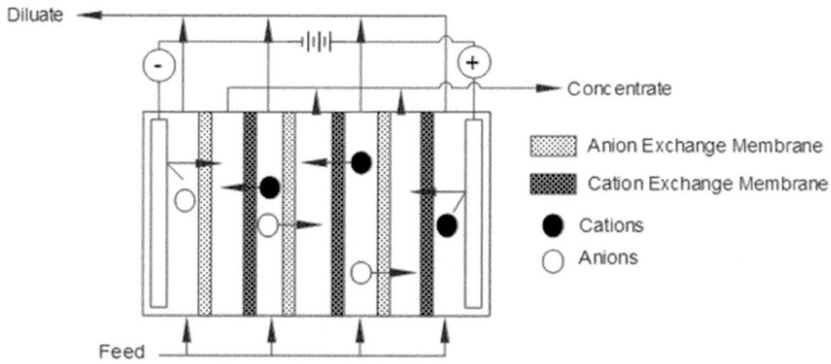

Fig. 3.1: Schematic of electrodialysis process

Feed enters the chambers between the membranes, where ions are removed via the ion exchange membranes. The diluent stream (low ionic concentration, "Diluate") and the concentrate stream (high ionic concentration) are withdrawn. Ions pass through the membrane based on the principle of electrolysis. Electrolytes are placed between two electrodes: the cathode (attracts cations, reduction) and the anode (attracts anions, oxidation), resulting in a redox reaction (equation 3.1). In electrodialysis, two electrodes are connected to a DC source. When switched on, the current dissociates the feed into cations and anions, which are then attracted to ion selective membranes between the electrodes, separating the ions.

$$\left.\begin{aligned} &M_mN_n + mM^+ + nN^{(m/n)-} \\ &C: mM^+ + me^- \rightarrow mM \\ &A: nN^{(m/n)-} - me^- \rightarrow nN \end{aligned}\right\} \tag{3.1}$$

Before delving into electrodialysis, we need to understand electrochemistry. The following sections provide a brief overview of electrolysis.

3.2 Understanding Electrolysis and Its Relevance to Electrodialysis?

Electrolysis dissociates ionic liquids using electric current. It requires a conductive solvent, two electrodes, and a DC source. When switched on, the current dissociates the solvent into ions. Positive ions (cations) move to the cathode and negative ions (anions) to the anode. At the cathode, cations gain electrons and are neutralized, while at the anode, anions release electrons. The cathode corrodes and is known as the "sacrificial electrode" (equation 3.1).

The choice of cathode and anode depends on redox potential. Anode should have a lower redox potential than the cathode, e.g., Cu as anode and Ag as cathode. Understanding the right electrode combination is crucial for effective ion separation during electrodialysis. Consider a case study on hydrogen production using a polyelectrolytic membrane (PEM) (21). Water is fed into a PEM unit for electrolysis, producing hydrogen and oxygen. Electrolysis requires 237.2 kJ/mol of electrical energy and 48.6 kJ/mol of thermal energy (22). Practically, the electrolyzer operates between 1.6 V and 2 V to split water. Figure 3.2 shows a schematic of the water electrolyzer setup.

$$\mathrm{H_2O} \xrightarrow[\text{48.6 kJ/mol thermal}]{\text{237.2 kJ/mol electrical}} \mathrm{H_2} + \frac{1}{2}\mathrm{O_2} \tag{3.2}$$

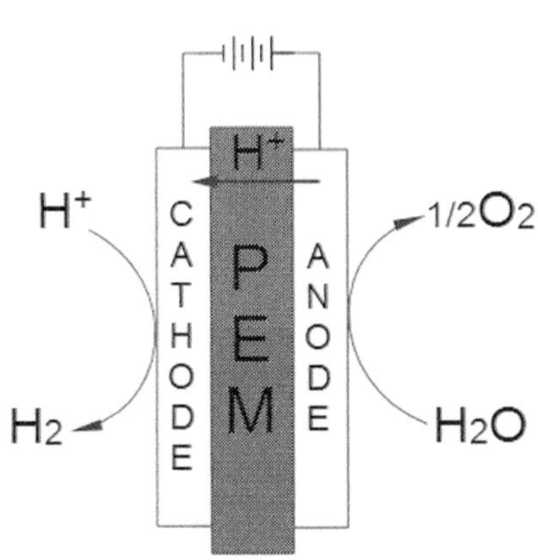

Fig. 3.2: PEM water electrolyzer

The figure depicts a Proton exchange membrane (PEM) electrolyzer, where the solid membrane (polysulfonated membrane, also called Nafion membrane) is used as the electrolyte. Pt ($E^0_{Pt^{2+}/Pt} = 1.18\,\text{V}$)/Pd ($E^0_{Pd^{2+}/Pd} = 0.915\,\text{V}$) is the cathode and IrO_2 ($E^0_{IrO_2/Ir} = -0.73\,\text{V}$)/RuO_2 ($E^0_{RuO_2/Ru} = -0.8\text{V}$) is the anode with a high current density of around 2Acm^{-2} at a low temperature range of 20–80°C (23). Selecting the right electrodes is crucial for achieving ion affinity through the membrane. The following sections will delve into electrodialysis and design considerations.

3.3 Classification of Electrodialysis

It is clear that electrodialysis combines electrolysis and ion exchange membranes (IEM). Figure 3.3 shows a schematic for proton exchange in Nafion membranes.

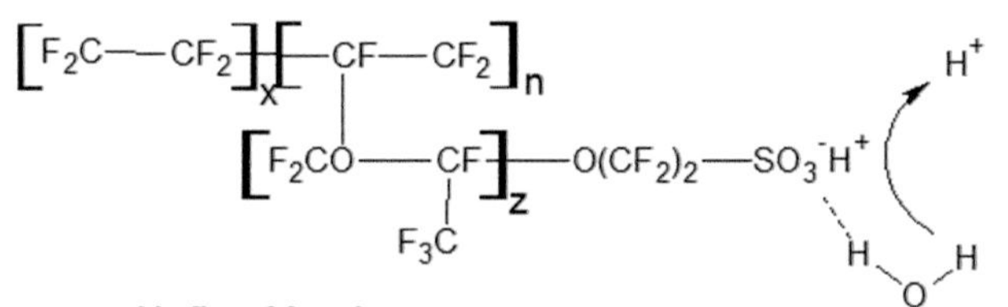

Fig. 3.3: Proton transport across Nafion membrane

Hence, it is evident that membranes play a crucial role in electrodialysis, as ion transport depends on membrane affinity for ions and the process itself. Electrodialysis can be categorized into (a) conventional, (b) bipolar membrane, (c) reverse, (d) selective, (e) two-phase and (f) electro-deionization electrodialysis. Conventional electrodialysis has already been explained, and details of the other processes are described in the following subsections.

3.3.1 Bipolar Membrane Electrodialysis (BMED)

This process involves a membrane with two polymer layers: one for anions and one for cations. Before elaborating on BMED, let's introduce two essential terms for the electrodialysis unit or electrolysis process. The "anolyte" is the electrolytic solution at the anode, and the "catholyte" is the electrolytic solution at the cathode. Before discussing the application of these electrolytes, let's understand how the bipolar membrane works. Bipolar membranes have two layers – anion exchange and cation exchange. Between these layers is an interfacial layer where water dissociates under an electric field. The interfacial layer is crucial for water dissociation or splitting. It may be impregnated with a catalyst to facilitate dissociation at lower energy. In this layer, water dissociates with H^+ moving towards the cation exchange membrane and OH^- towards the anion exchange membrane (see Figure 3.4). For H^+ ions to cross the cation exchange layer, it must face the cathode; the opposite applies for the anion exchange layer. Facing the cation exchange membrane towards cathode and anion exchange membrane towards the anode is called "reverse bias". When the polarity is inverted, it is called forward bias. In reverse bias, ions move away from the interfacial layer, increasing the electrical resistivity of the membrane. Once a certain voltage is reached, dissociation begins, depleting the interfacial layer and showing maximum resistance. This is followed by a rapid increase in current due to water dissociation. Figure 3.5 shows a current versus open circuit voltage (OCV) curve illustrating ions transport. After a certain voltage, current may not increase current due to water depletion in the interfacial layer. In forward bias H^+ and OH^-, move towards the junction and recombine into water, making the matrix electrically conductive.

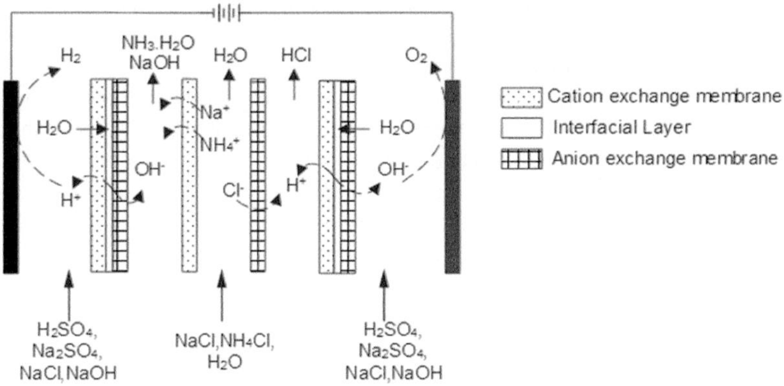

Fig. 3.4: BMED for desalination of water

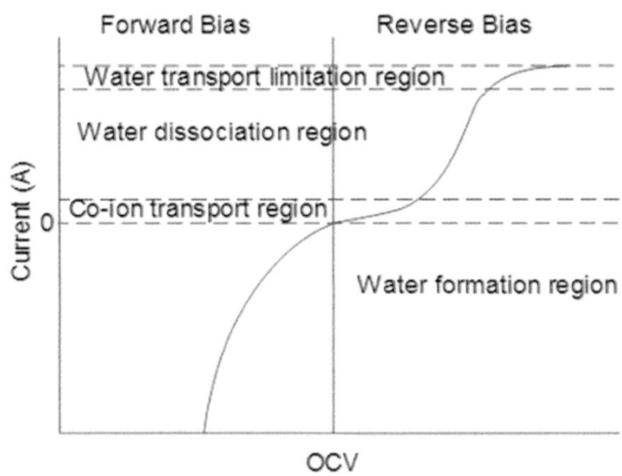

Fig. 3.5: I-V curve for bipolar membrane

3.3.2 Reverse Electrodialysis (RED)

Electrodialysis, contrary to its conventional form, applies voltage against the membrane through electrodes and electrolytes. The electrical potential transports ions across the membrane, achieving purification. In RED, the salinity gradient creates a potential, and the chemical potential difference drives ion transport across the membrane generating electrical current. Thus, RED can be considered a battery unit generating electrical power. Two electrodes, a cathode and an anode, are placed on either side of cation (typically with fixed charge SO_3^-) and anion (typically with fixed charge NR_3^+) exchange membranes (24). Figure 3.6 depicts a schematic for RED. Introduced in the 1950's-1960's, RED initially powered electrodialysis units. By the 1970, RED was recognized as a potential power generator, capable of producing around 2.6 TW (25).

A key consideration in the RED process is the electrode system. An inert electrode system, such as mercury or carbon, doesn't participate in chemical reactions with circulating electrolytes. However, chemical reactions either may or may not occur on the electrodes. $FeCl_2/FeCl_3$ is used as the catholyte and $K_4Fe(CN)_6/K_3Fe(CN)_6$ as the anolyte in a system where inert electrodes experience no reactions. However, the main drawback is the stability and pH acceptability of these electrolytes. With reactions occurring on the inert electrode, Na_2SO_4 is used as the catholyte and NaCl as the anolyte, resulting in a voltage loss of around 2-3 V. Participating electrodes are involved in the redox reactions. Participating electrodes include Ag/AgCl with circulating NaCl, Cu with circulating Na_2SO_4, Zn with circulating $ZnCl_2$, $Na_{2-x}Mn_5O_{10}$ for battery systems. For capacitive electrodes, activated carbon with circulating NaCl is used in RED units (25). Salinity gradient power (SGP) uses RED for power generation,utilizing solutions with maximum salt concentration differences. A recent ELGALabWater, USA study on SGP generation analyzed multivalent ions in brackish water (26). The study found that power density is significantly reduced when multivalent ions are present. To maintain moderate power generation in a RED-based unit for SGP water softening is recommended.

Donnan exclusion is the phenomenon responsible for permeating a specific charged ions through the membrane. Now, what is this? The membrane's fixed charge group has an affinity for counter ions. For example, the $-SO_3^-$ group in cationic exchange membranes has an affinity for Na^+ ion (after softening). Na^+ is attached to six water molecules, wetting the membrane and creating two phases: the polymer layer (the membrane) and the gel layer (fixed charges with counter ions). The gel layer is free of any co-ions due to the negative fixed charge (Figure 3.7). This is called Donnan exclusion. The theory extends to removing undesired ions (e.g., in brackish water) before feeding to a RED unit, enhancing efficiency. For instance, high levels of Mg^{2+} with Na^+ in water resistance to power generation by nearly 75% in a RED unit. By contacting the feed with a cation exchange membrane and providing high Na^+ concentration on the other side, ion transport occurs. Mg^{2+} moves to the permeate side, while Na^+ enriches the feed, allowing for efficient RED. This is known as Donnan dialysis in conjunction with RED.

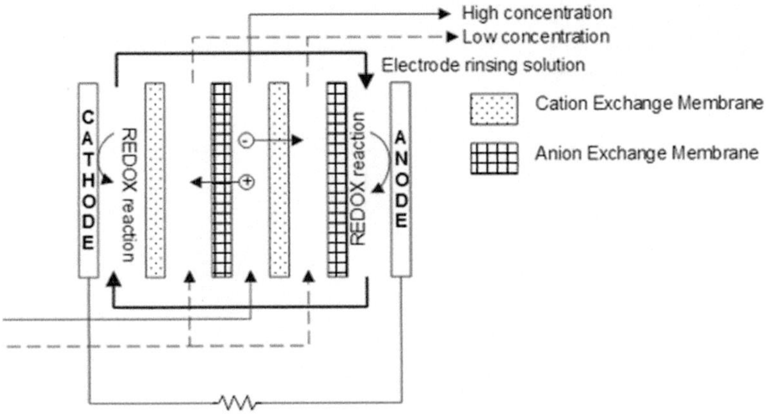

Fig. 3.6: RED schematic

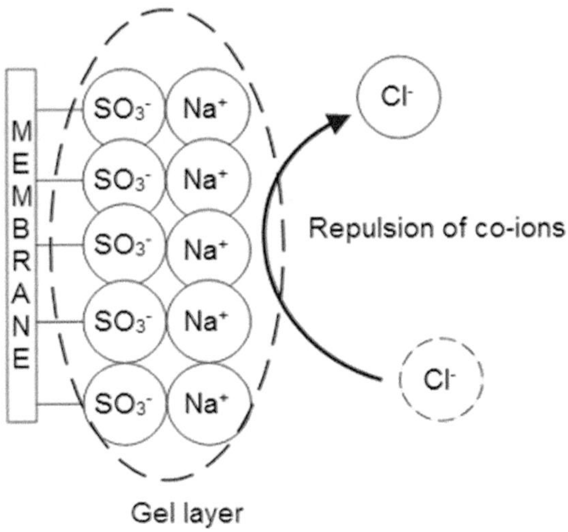

Fig. 3.7: Donnan exclusion principle

3.3.3 Two-phase Electro-electrodialysis

Before delving into two phase electro-electrodialysis, note that it sometimes occurs with a conventional electrolysis cell. As shown in Figure 3.4, hydrogen evolves at the electrode and acts as the dispersed phase at the electrode-electrolyte interface. The evolution of the gaseous phase on the electrode creates a shielding effect (27). Before analyzing two-phase electro-electrodialysis, let's understand two phase electro-electrodialysis and its difference from electrodialysis. The main difference is the reformation of compounds from ions (Figure 3.8). Unlike conventional electrodialysis, two-phase electro-electrodialysis separates organic and aqueous phases using the electrodialysis technique. Figure 3.8 shows a schematic of this process.

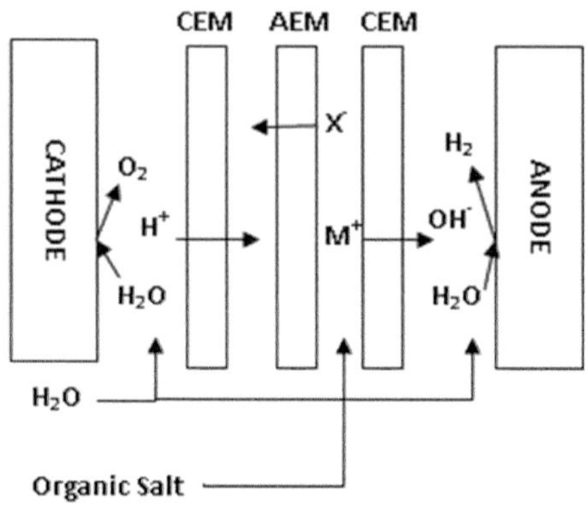

Fig. 3.8: Schematic of two phase electro-electrodialysis (AEM: Anion exchange membrane; CEM: Cation exchange membrane)

Organic salt (organic phase) and water are fed into channels set by membranes and electrodes. Water serves as the anolyte and catholyte against the electrodes. Equations 3.3 and 3.4 show the cathodic and anodic reactions respectively.

$$\text{Cathodic reaction: } \begin{cases} H_2O \rightarrow H^+ + OH^- \\ 2H^+ + 2e \rightarrow H_2 \uparrow \end{cases} \tag{3.3}$$

$$\text{Anodic reaction: } \begin{cases} H_2O \rightarrow H^+ + OH^- \\ 2OH^- - 2e \rightarrow H_2O + \frac{1}{2}O_2 \uparrow \end{cases} \tag{3.4}$$

In Figure 3.8, an anion exchange membrane is between two cation exchange membranes. During electrolysis, anions pass through the anion exchange membrane and cations through the cation exchange membrane. H^+ and anions combine to form organic acid. Electro-osmotic pressure in electro-electrodialysis restricts organic acid formation. The main disadvantages are low product recovery due to low selectivity and membrane stability issues (28).

3.3.4 Electro-deionization Electrodialysis

The process combines ion exchange resin and electrodialysis. Figure 3.9 shows a schematic. The system features two electrodes adjacent to the ion exchange membranes, with a mixed resin bed in between. Wastewater with high TDS is fed through this setup. High TDS wastewater is fed into every chamber as shown in Figure 3.9. During its travel through the resin bed, ions are entrapped and percolate through either cation or anion exchange membranes, depending on their charges due to the applied electrical potential. The chamber filled with resins is called the dilute chamber, while the chambers collecting ions are called the concentrate chambers.

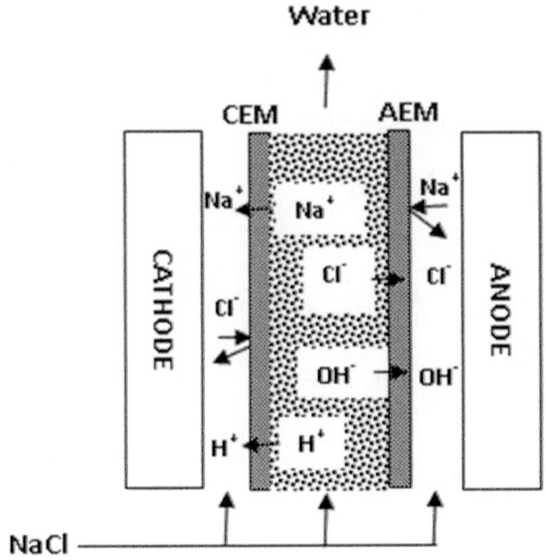

Fig. 3.9: Schematic of electro-deionization electrodialysis (AEM: Anion exchange membrane; CEM: Cation exchange membrane)

Cation exchange resins (R–C) and anion exchange resins (R–A) consist of polymeric resins (R) with attached ions. Resins are categorized as either strong or weak based on electrolytes. Strong cation exchange resins have sulfonic or phosphoric groups, while phenolic or carboxylic groups indicate weak resins (29). Strong anionic resins have ammonium base polymer and hydroxide groups, while weak resins have primary and secondary amines. Counterions replace mobile ions and adsorb on the surface. When electrical potential is applied, adsorbed ions detach and move through membranes. A key advantage is resin bed regeneration and ion separation. However, pretreatment to remove the suspended solids is essential to avoid clogging and pressure drop (28). Khan et al. (29) note that resin with narrow particle size enhances ion transport.

The discussion so far on electrodialysis covers its operational sequences and modes. Membranes are crucial for the process's effectiveness. Charge affinity towards the membrane is essential for ion separation from the solvent. In electro-deionization electrodialysis, compromised membrane affinity hinders resin bed regeneration, leading to saturation with no further separation. Efficiently charged membranes ensure effective electrodialysis. Typically, membranes used are copolymers of styrene and divinylbenzene.

3.4 Mathematical Interpretation of Electrodialysis

Figure 3.10 shows a control volume around the electrodialysis membrane with a solute balance. Two main assumptions for the mathematical model are: no water permeation through membranes and the formation of a thin mass transfer boundary. The solute balance over the control volume gives,

$$u\Delta x WC\big|_{x} - u\Delta x WC\big|_{x+\Delta x} = J_{W,A} W\Delta y + J_{W,C} W\Delta y \tag{3.5}$$

$$\therefore u\frac{\partial C}{\partial y} = -\frac{\partial J_{W,A}}{\partial x} - \frac{\partial J_{W,C}}{\partial x} \tag{3.6}$$

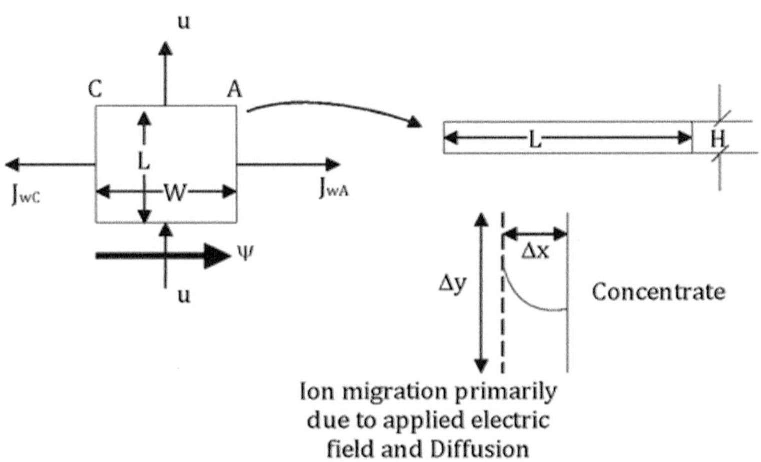

Fig. 3.10: Control volume during separation of ions (A: Anionic membrane; C: Cationic membrane)

Equation 3.6 doesn't imply the flux because of the molecular diffusion only, but also because of the externally applied electrical field 'ψ'. Now the flux term J_W in terms of chemical potential can be given by equation 3.7.

$$J_w = -\frac{DC}{RT}\frac{d\mu}{dx} \tag{3.7}$$

where 'C' is the solute concentration, 'μ' is the chemical potential of the solutes, 'T' is the temperature, and D is the diffusivity of the solutes. Due to the applied electrical field facilitating ion transfer, chemical potential can be replaced by electrical potential as shown in equation 3.8.

$$J_w = -\frac{DC}{RT}\frac{d\eta}{dx} \tag{3.8}$$

Further, electrical potential (η) can be replaced by $\eta + zF\psi$ as shown in equation 3.9.

$$\eta = \mu + zF\psi \tag{3.9}$$

where, '*z*' is the charge number, '*F*' is Faraday's constant and ψ is the electrical potential. From equations 3.8 and 3.9, the final form of the flux equation is,

$$J_w = -\frac{DC}{RT}\frac{d}{dx}(\mu + zF\psi) = -\frac{DC}{RT}\frac{d\mu}{dx} - \frac{DC}{RT}zF\frac{d\psi}{dx} \tag{3.10}$$

Further, considering the electrolysis of M_mN_n, equation 3.10 can be rewritten separately for cations and anions (equation 3.11).

$$\left.\begin{aligned} J_{w,A} &= -\frac{D_A nC}{RT}\frac{d\mu_A}{dx} - \frac{D_A Z_A nC}{RT}F\frac{d\psi}{dx} \\ J_{w,C} &= -\frac{D_C mC}{RT}\frac{d\mu_C}{dx} - \frac{D_C Z_c mC}{RT}F\frac{d\psi}{dx} \end{aligned}\right\} \tag{3.11}$$

Local current density (I_L) due to ions is given by equation 3.13.

$$\left.\begin{aligned} I_A &= Z_A F J_{w,A} \\ I_C &= Z_C F J_{w,C} \end{aligned}\right\} \tag{3.12}$$

In the absence of an applied electric field there will be no net current, yielding,

$$I_A + I_C = Z_A F J_{w,A} + Z_C F J_{w,C} = 0 \tag{3.13}$$

From equation 3.11 and 3.13,

$$F\frac{D_A nC}{RT}\left(-\frac{d\mu_A}{dx} - Z_A F\frac{d\psi}{dx}\right) + mF\frac{D_C mC}{RT}\left(-\frac{d\mu_C}{dx} - Z_C F\frac{d\psi}{dx}\right) = 0 \tag{3.14}$$

$$\Rightarrow nD_A Z_A\left(-\frac{d\mu_A}{dx} - Z_A F\frac{d\psi}{dx}\right) + mD_C Z_C\left(-\frac{d\mu_C}{dx} - Z_C F\frac{d\psi}{dx}\right) = 0$$

$$\Rightarrow \frac{d\psi}{dx} = -\frac{nD_A Z_A\frac{d\mu_A}{dx} + mD_C Z_C\frac{d\mu_C}{dx}}{nD_A F Z_A^2 + mD_C F Z_C^2}$$

Combining equation 3.11 and 3.14,

$$\left.\begin{aligned} J_{w,A} &= -\frac{D_A nC}{RT}\left[\frac{d\mu_A}{dx} - Z_A F\frac{nD_A Z_A\frac{d\mu_A}{dx} + mD_C Z_C\frac{d\mu_C}{dx}}{nD_A F Z_A^2 + mD_C F Z_C^2}\right] \\ J_{w,C} &= -\frac{D_C mC}{RT}\left[\frac{d\mu_C}{dx} - Z_c F\frac{nD_A Z_A\frac{d\mu_A}{dx} + mD_C Z_C\frac{d\mu_C}{dx}}{nD_A F Z_A^2 + mD_C F Z_C^2}\right] \end{aligned}\right\} \tag{3.15}$$

Therefore, flux prediction for the ions through ionic membranes can be approximated with equation 3.15 once their chemical potential is known.

3.5 Membranes Used in Electrodialysis and Fouling

Several membranes have been studied so far for electrodialysis. In 2022, Titorova et al. (30) used two sulfonated cation exchange membranes (CEM) Neosepta CMX and CJMC-5. The matrix of the CMX membrane was polystyrene-divinylbenzene with an electrical resistance of 2.5 to 3.5 ^.cm^2 and an ion exchange capacity of 1.69 meq per g (31). CMX membrane is denser than CJMCEM, with the lowest water content at 8 mol per mol of functional groups. In contrast, CJM-5 has 23 and CJM-3 has 27 mol of water per mol of functional groups. In general, CEM is generally divided into homogeneous and heterogeneous groups. Homogeneous membranes are typically produced using the "paste" method and fabricated at the nanoscale scale, up to 100 nm (32). CMX membrane is from Neosepta, with several other commercial membranes listed in Table 3.1 for ion exchange.

In membrane separation processes, fouling occurs due to solute deposition on the membrane surface by charged ions. If solutes detach and return to the solution, fouling can be removed, restoring the membrane. The process is self-cleaning through polarity reversal, repelling co-ions, known as electrodialysis reversal (EDR). Various foulants can form on the membrane surface during electrodialysis. A key concern is scaling fouling from Ca^{2+} and Mg^{2+} in hard water leading to nucleation and hard crystal formation on the membrane. In the EDR process, ion depletion in the diluant chamber increases resistance to current. Water splitting happens at the CEM produces hydroxyl ions (OH^-) and H^+ ions. OH^- ions pass through the CEM (hydroxyl leakage), dissolving metal ions to form partially soluble metal hydroxide (34), leading to membrane fouling.

Colloidal silica in water can form a fouling layer on the AEM due to its negatively charged surface.This deposition blocks channels, reducing membrane exchange efficiency. Silica and organic matter (NOM) deposition on AEM can cause irreversible fouling, resistant to chemical removal. Factors influencing NOM fouling include pH, solubility, molecular charge and structure. For example, lignin with phenolic functional groups undergoes protonation, forming lignin colloids through self-aggregation. Van Der-Waals forces create a deep fouling layer on the surface of the membrane (34). The presence of NOM along with Ca^{2+} ions leads to stable fouling layer formation via Ca^{2+}-NOM complexation. This reduces Ca^{2+} charge, complicating membrane cleaning with EDR (35). pH determines membrane fouling stability. The isoelectric point (net surface charge is zero) serves as the pH threshold. If pH is below the isoelectric point, CEM has a net positive charge; if above, AEM has a net negative charge. The complex forms a stable fouling layer on the membrane through simple adhesion (35). During EDR, polarity reversal transfers foulants from the concentrate side to the diluant side, as in electrodialysis. EDR is essential in electrodialysis to mitigate membrane fouling and ensure process sustainability. Figure 3.11 shows the EDR process. After polarity reversal, ions in the diluant zone near the electrodes deplete,

Table 3.1: Commercial ion exchange membranes available (33)

Membrane fabrication house	Commercial name	Ion-exchange capacity (m eq per g)	Thickness of the membrane (μm)	Permselectivity (%)	Membrane areal resistance (∧cm^2)	Nature of the membrane
Fumasep	FKE	1.36	34	98.6	2.46	CEM, Hmg
	FKD	1.14	113	89.5	2.14	CEM, Hmg
	FKS	1.54	40	94.2	1.5	CEM, Hmg
Fujifilm	CEM RP1 80050-04	1.45	120	90	2.96	CEM, Hmg
Neosepta	CM-1	2.30	133	97.2	1.67	CEM, Hmg
	CMX	1.62	164	99	2.91	CEM, Hmg
	CMS	2.0	150	-	1.5-2.5	CEM, Hmg
Nafion	117	0.9	200	97	1.5	CEM, Hmg
	901	1.1	400	96	3.8	CEM, Hmg
RAI	R-5010-H	0.9	240	95	8-12	CEM, Hmg
	R-1010	1.2	100	86	0.2-0.4	CEM, Hmg
Selemion	CMV	2.01	101	98.8	2.29	CEM, Hmg
Qianqiu	CEM	2.01	205	82	1.97	CEM, Hmg
CSMCRI	IPC	1.4	140-160	97	1.5-2	CEM, Hmg
	HGC	0.67-0.77	220-250	87	4-6	CEM, Hetero
Ralex	CMH-PES	2.34	764	94.7	11.33	CEM, Hetero
	CM	2.15	-	84	9.36	CEM, Hetero
Selemion	DSV	1.89	121	88.9	1.03	AEM, Hmg
	APS	0.29	138	88.4	0.68	AEM, Hmg
	AMV	1.78	124	87.3	3.15	AEM, Hmg
Ralex	AMH-PES	1.97	714	89.3	7.66	AEM, Hetero

Hmg: Homogeneous; Hetero: Heterogeneous

restricting electrical conductance and leading to the parameter "limiting current density" (equation 3.16) (36).

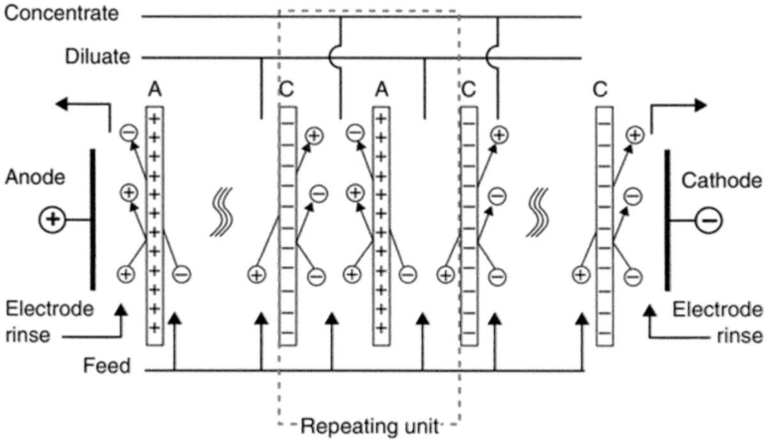

Fig. 3.11: Schematic of EDR process

$$i_{lim} = \frac{C_{feed} D_i ZF}{\delta\left(t_{im} - t_{is}\right)} \tag{3.16}$$

The term, the ratio between diffusivity and the membrane thickness, is called the mass transfer coefficient and is correlated by equations 3.17 and 3.18.

$$= \frac{D_i}{} = 1.62\left(\frac{uD_i}{d\ L}\right)^{0.33} \quad \left(\text{Laminar; Re} < 2100\right) \tag{3.17}$$

$$k = \frac{D_i}{\delta} = 0.023 \frac{u^{0.8} D_i^{0.672}}{d_h^{0.2} \nu^{0.47}} \quad \left(\text{Turbulent; Re} \geq 2100\right) \tag{3.18}$$

EDR must be considered with electrodialysis, forming the EDR process. However, key disadvantages include decreased electrical conductance near the electrode due to ion depletion water splitting causing insoluble metallic salt deposition, leading to fouling. Another disadvantage is the substantial capital cost. To reduce costs, the integrated EDR-RO process is considered.

3.6 Integrated EDR-RO Process for Oily Wastewater

Oily wastewater, containing fats, oils and greases (FOG) from industrial and domestic sources, is difficult to treat. Membrane separation faces challenges due to fouling from oil layers on the surface. Designing an efficient membrane separation process is challenging and extensively researched. The main issue with oily wastewater is the the difficult-to-separate oil-water emulsion. The most effective method to break the emulsion and separate oil and water, is the electrolytic approach. EDR separates ions and breaks emulsions. This process will be extended to ion separation using RO to achieve "Zero Liquid Discharge (ZLD)." A case study by Venzke et al. (37) involved pretreating oily wastewater with a sand filter, followed by an activated charcoal bed. The feed goes to an RO unit, then the permeate to an EDR unit. Permeate and the diluant from EDR were merged, recovering 87.3% of ultrapure water. In another study by the same researcher (38), EDR followed by RO achieved around 90% water recovery. Electrolyte loading can cause deposition and fouling on the RO membrane. Applying EDR before RO reduces ion load, potentially increasing recovery. The EDR-RO integrated system shows 85-99% pollutant mitigation efficiency, while RO-EDR shows 70-95% efficiency. Literature on the integrated process is limited, but its main application is in brackish water treatment for potable water recovery.

3.7 Integrate EDR-RO System for Brackish Water Treatment

Desalination of brackish water compensates for depleted potable water sources. Membrane distillation and RO are key technologies, with RO being the most conventional for potable water recovery. The nascent integrated EDR-RO technology feeds RO reject to the EDR unit. Literature is limited, but EDR removes 94% TDS at 27°C and 30,000 ppm TDS (39). Applied voltage, around 18 V (Far et al., 39), is crucial for TDS removal. EDR effectively removes hardness, crystallizing and disposing of SO_4^{2-} and CO_3^{2-} (40). The integrated RO-EDR system achieves around 91.6% recovery, with EDR treating RO reject. The Langelier Saturation Index (LSI) is an important parameter for measuring water corrosion and scale formation. LSI is measured by the difference between the system pH and the equilibrium pH (H^+ and CO_3^{2-} in equilibrium). A positive LSI indicates excess CO_3^{2-} ions, leading to scale formation. A negative LSI indicates more H^+ ions, leading to aggressive corrosion. According to Turek et al. (40), the RO-EDR technology can recover 91.6% water even at an LSI of 2.29 indicating aggressive scaling.

3.8 Summary

- Electrodialysis in the light of electrolysis.
- Different technologies with electrodialysis.
- Membranes used in electrodialysis and membrane fouling.
- Self-cleaning technology EDR.
- Integrated EDR-RO technology for the treatment of oily wastewater and brackish water.

CHAPTER

4

Gas Separation and Pervaporation

4.1 Introduction

In both gas separation and pervaporation, partial pressure differences between the feed and permeate sides drive the separation process. During gas separation, high pressure is often developed at the feed side using a compressor, while atmospheric pressure is maintained on the permeate side (see Figure 4.1(a). In pervaporation and vapor permeation, the feed side is mostly at atmospheric pressure while a reduced pressure zone is developed at the permeate side by using a hybrid system of a condenser and a vacuum pump (see Figure 4.1(b). Transport phenomenon in these membrane separation processes are described by the well-acknowledged solution-diffusion model, generally following three consecutive steps as outlined below:

- Selective sorption of the preferentially permeating component into the membrane layer from the feed stream.
- Selective diffusion of the identical component across the membrane thickness.
- Desorption of permeate stream at the permeate side from the downstream membrane surface.

Among these processes, pervaporation is comparatively more complex than the other two processes. This is because a phase transition is involved in pervaporation in which the latent heat of vaporization of the preferentially permeated components must be provided from outside. This signifies that in addition to mass transfer, heat transfer also takes place. Pervaporation is sometimes compared with the distillation process because of the presence of liquid phase and vapor phase in the feed and permeate sides respectively. However, the separation mechanism is completely distinct for these two processes. The prime importance of pervaporation is that unlike distillation, the separation is not dependent on the thermodynamic equilibrium between the liquid and vapor phase. This implies that the permeate concentration is not controlled by the vapor-liquid equilibrium (VLE) but by the preferential permeability of the components across the membrane, which is eventually dependent on the component's solubility and diffusion rate.

Hence, pervaporation is considered as a green alternative technology to distillation process especially in scenarios where energy demand is exceedingly high because of the presence of azeotropes (close boiling point components).

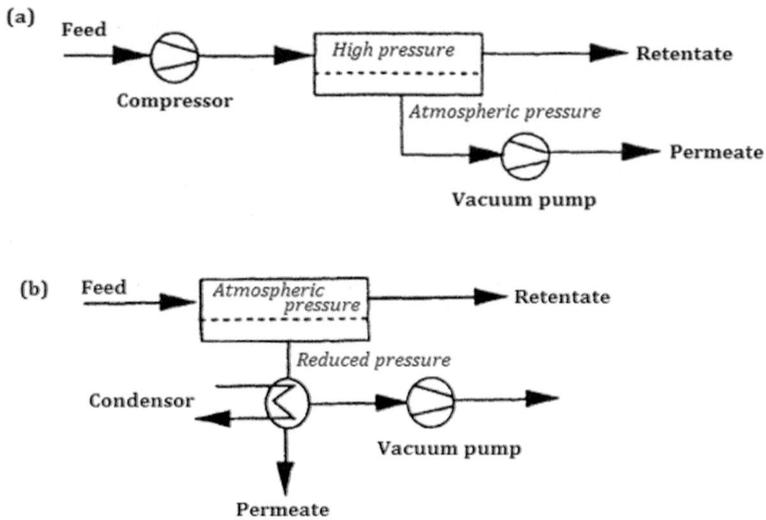

Fig. 4.1: Schematic representation of (a) gas separation process and (b) pervaporation process

4.1.1 Historical Perspective

4.1.1.1 Gas Separation

Mitchell's 1831 discovery showed gases penetrate natural rubbers at different rates (41). Thomas Graham later proposed the solution-diffusion mechanism for gas transport through membranes. Experiments on membrane based gas separation for oxygen began in the 1860s. The first attempts to synthesize the composite membrane materials as and observe hydrogen permeation through inorganic membranes were made then (42, 43). In the 1930s and 1940s, R.M. Barrer conducted significant gas permeation studies and derived the widely-used permeability coefficient unit. The unit was defined as 1 Barrer = 10^{-10} cm^3 (STP). cm/(cm^2.s.cmHg) and the corresponding SI unit is mol.m/m^2.s.Pa (1 Barrer = 3.35×10^{-16} mol.m/m^2.s.Pa). Until 1945, the industrial application of the gas separation process involving microporous membranes was uncommon. The first major initiation in this context was the use of inorganic membranes for separating uranium isotopes for nuclear and military purposes (44). Despite their higher selectivity compared to microporous membranes, the high cost of non-porous polymeric membranes limited their use in gas separation processes until the 1970s. Initially, membrane thickness caused unsatisfactory lower flux rates, limiting their commercial use. In 1963, Loeb and Sourirajan developed integrally skinned asymmetric membranes, solving this issue. Modifications for gas separation maintained the integrity between the porous and thin dense-selective layers. Initially developed as flat sheets, asymmetric membrane structures evolved into hollow fibers (45). The Loeb-Sourirajan technique significantly advanced MF, UF, NF, and RO membranes for liquid separation and facilitated membrane commercialization for gas separation.

In the 1980s, membrane technology in gas separation advanced significantly. Henis and Tripodi's revolutionary invention of composite membranes, with a thin film of silicone-rubber on an asymmetric substrate, achieved high hydrogen selectivity at low cost (46). This breakthrough led to the commercialization of Monsanto Co.'s Prism system in 1980 for recovering hydrogen from ammonia plant purge gas (44). The success of Monsanto Company has extensively stimulated the urge of intensive research related to development of novel membranes and gas separation processes. In successive years, lewis acid-base complex dopes were introduced by the membrane engineers in Permea for synthesizing asymmetric membranes in the form of hollow fibers with graded-density skins (47, 48). The Dow Company developed membrane based systems for the separation of nitrogen from air, and Cynara and Separex Company for separating carbon dioxide gas from the natural gas. In 1990, another remarkable invention by Pinnau and Koros was the development of defect-free ultrathin asymmetric membranes using a dry/wet phase inversion method. This elevated the gas permeation rate across the membrane substantially to fulfill the requirement for industrial large-scale applications (49). Later on, most of the research studies were focused primarily on synthesis of the highly performing asymmetric hollow fibers. For instance, Chung et al. successfully tuned Permea's technology for fabricating 6FDA-polyimide membranes to carry out air separation (42).

A simple setup used for membrane based gas separation process is not substantially distinguishable from that being utilized for the separation of solid–liquid except the fact of distinct phases in the streams. The gas mixture is introduced within the membrane module at a very high pressure in which the diffusion of one component occurs faster across the membrane and the permeate stream is eventually enriched with that component. The rest of the gas phase is concentrated and collected as the residue stream. For attaining more complete separation process, recycling of some fraction of the permeate stream or residue stream is essential.

4.1.1.2 Pervaporation

The origin of pervaporation had occurred in the early nineteenth century. In 1906, Kahlenberg was the first to utilize a very thin rubber sheet for carrying out the selective separation of alcohol and hydrocarbons (50). However, the term "pervaporation" was introduced in 1917 by Kober who combined the membrane permeation process with distillation, crystallization, and evaporation for the first time. Initially, cellulose acetate collodion bags were utilized for permitting evaporation of water at a rapid rate from the aqueous solution mixture. Later in 1935, the pervaporation technique was used for concentrating the solution consisting of proteins (51). In 1942, Melnick adopted a hybrid process involving both pervaporation and ultracentrifugation to concentrate the dilute solutions of the Theiler's murine encephalomyelitis virus (TMEV) (52). Subsequently, a patent was filed by Binning and Stuckey in 1954 for using the pervaporation process with an ethyl cellulose permselective membrane for separating hydrocarbons (53). This team also explored the separation of aromatics by pervaporation in 1958 (54-56). Later on, in 1962, they also filed a process patent on the dehydration of organic chemicals by pervaporation (57). Although PV provided good separation, it was not acceptable to process industries due to low process flux.

Furthermore, the dehydration of alcohols like ethanol from aqueous mixtures using regenerated cellulose membranes was explored (58). It was eventually revealed that the driving force, resulting from the pressure gradient through the sorption of specific components on the pervaporation membrane surface, promotes the transport of solutes (50). However, despite providing a satisfactory separation factor, the pervaporation process faced the intricate challenge of lower permeation flux rate, limiting its economic compatibility. Many years later in 1982, the dehydration of an aqueous-ethanol mixture was successfully accomplished using a novel thin layer of poly vinyl acetate (PVA) crosslinked onto a porous polyacrylonitrile (PAN) membrane.

In the field of pervaporation, a major breakthrough was the 1960s invention of the phase inversion method for membrane casting by Loeb and Sourirajan. This process enabled the synthesis of a thin active layer on the porous asymmetric support, elevating the permeation flux (59). In 1970, Charles studied the separation or removal of volatile pollutants using selective hollow fibers. He then analyzed the effect of different operating parameters on enhancing ammonia flux and assessed its feasibility. In 1972, Aptel studied the separation of various liquid mixtures, such as hydrocarbons-chloroform or ethyl ether, alcohol-water, and water-nucleophilic organic solvents, using polytetrafluoroethylene membrane (60). Consequently, the pervaporation process utilizing polytetrafluoroethylene films was explored in 1976 for separating azeotropic mixtures (61). During the timeline of 1982 to 1996, several pervaporation pilot plants were successfully installed in varied process industries. In 1982, GFT Membrane Systems of Germany installed a pervaporation pilot plant to enrich ethanol with an azeotropic mixture of ethanol-water. Subsequently, Separex, USA installed a pervaporation pilot plant in 1988 for separating methanol from a mixture of methyl tert-butyl ether and isobutene (62). Then in 1996, Membrane Technology and Research Inc., U.S.A., indigenously synthesized a membrane and made a major breakthrough by installing a commercial pervaporation plant for removing trace quantities of volatile organic compounds (VOCs) from industrial effluent (63). From 1990 onwards, research studies were focused on the modifications of membranes for improving selectivity, permeation flux, and membrane strength. In the pervaporation process, chemical compatibility of targeted species and membrane materials has a significant influence on efficiency. Another key factor affecting process performance is the variation in diffusivity of different feed components. During this period, several novel membranes, such as polyolefin/polydimethylsiloxane composite membranes and silica membranes, were developed and studied for pervaporation (64). Between 1995 and 2000, research initiatives aimed to improve the hydrophilicity of membranes used in the pervaporation process. Notably, this led to the fabrication of zeolite-incorporated membranes (65–67). Moreover, hybrid processes combining pervaporation and other separation techniques were explored, emphasizing their design, applications, and economical feasibility. In 1999, Bryan et al., utilized pervaporation membranes for methanol fuel-cells (68, 69).

Over the next ten years, the primary research focus was on solvent dehydration from aqueous solutions. Examples include the dehydration of acetone (70), ethanol (71), butanol (72), 2-butanol (73), *n*-hexane, and *n*-hexane/benzene mixtures. Another major research interest during this period was the separation of organic-organic mixtures using pervaporation processes. This included the separation of cyclohexane/benzene mixtures (74), sulfur removal from gasoline fuel (75), VOC removal from aqueous solutions (76), recovery of ethanol, butanol, and acetone from Clostridium acetobutylicum fed-batch reactors (77), ethanol recovery from corn fibers (78), and recovery of final targeted products from biomass fermentation processes (79). Consequently, researchers have extensively worked on the synthesis and development of mixed matrix membranes through incorporation of varied inorganic fillers like nanoparticles, carbon nanotubes (CNT), and zeolites within the polymeric matrix to perform intricate separations utilizing pervaporation processes.

In the last decade, most of the research was carried out for the development of membranes attributed with enhanced durability, strength, and selectivity along with better efficiency. Examples of such membranes are membranes consisting of graphene oxides, metal organic framework (MOFs) nanoparticles, and membranes derived from biomaterials. Amalgamated processes like coupled pervaporation and gas stripping were also studied to achieve enhanced product recovery along with better purity (80). Also, the use of pervaporation process for brine desalination was observed which manifested total salt rejection in case of the inclusion of graphene oxide within the pervaporation membranes (81).

4.2 Types of Pervaporation Process

Pervaporation can be classified into three types based on the driving force used for separation: vacuum pervaporation, thermo-pervaporation, and sweep-gas pervaporation. Vacuum pervaporation is the most commonly observed method today. In this mode, the feed side is maintained at atmospheric pressure while a vacuum is created on the permeate side to ensure a sufficient driving force

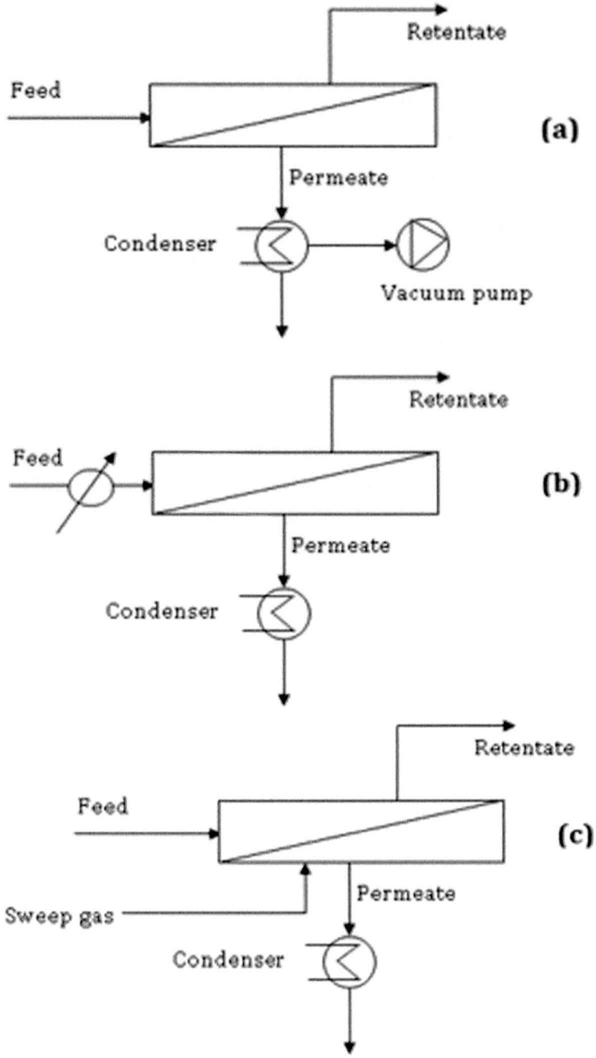

Fig. 4.2: Schematic representation of (a) vacuum pervaporation, (b) thermo-pervaporation, and (c) sweep gas pervaporation

required for effective separation (Figure 4.2(a)). Due to the vacuum condition, the permeate stream condenses and is collected as a liquid phase. Meanwhile, the retentate stream becomes concentrated with the less permeated components. The partial pressure gradient between the feed and permeate sides of the membrane acts as the driving force for separation. Due to varied diffusivities and solubilities of the components, different molecular transport rates across the membrane are observed, facilitating the separation selectivity. Moreover, thermo-pervaporation is an emerging process that develops significant driving force through a temperature gradient to separate the component of interest. To enhance the solutes' transport rate, the liquid mixture is generally preheated before being used as the feed stream, aiming to achieve attain a higher partial pressure in the thermo-pervaporation process (Figure 4.2(b)). The first step involves pre-heating the liquid mixture to approximately 60–80°C. At this elevated temperature, the partial pressure of the different feed stream components will increase. Thus, a sufficient driving force is produced for the easy transport of molecules, as the downstream permeate compartment is maintained at a much lower temperature as compared to the upstream feed compartment. This process benefits from the fact that the driving force is developed naturally, without the requirement of creating a vacuum condition. This approach becomes more promising if the temperature of the liquid feed stream reaches the aforementioned temperature range. Moreover, the thermo-pervaporation process benefits from higher mass transfer coefficients (MTC) at elevated temperatures, mitigating the issues of mass transfer loss and pressure drop in the feed channel. Another approach to the pervaporation process involves using sweeping gas as a purging gas on the permeate side to maintain the driving force for the solute transport through the membrane (Figure 4.2(c)). This method is commonly known as sweeping gas pervaporation or sweep gas pervaporation.

4.3 Principle of Separation of Multicomponent Mixture through Pervaporation

To ensure the liquid state of the feed stream during the pervaporation process, the feed solution is maintained at a temperature below its boiling point and at a pressure above its bubble point. The selection of the membrane is a crucial factor in the pervaporation process for achieving satisfactory separation efficiency, as it largely depends on the membrane's selectivity for the specified component. A dense membrane may be chosen for separation as it provides the necessary selectivity for the process. The permeated components of the mixture vaporize within the membrane until they reach the permeate side. Moreover, a vacuum condition is usually maintained on the permeate side of the membrane to ensure a large partial pressure gradient between the two sides. This gradient serves as the major driving force for separation. The permeated components in the permeate stream eventually condense and are recollected as the liquid phase. The liquid stream that is retained and has not permeated across the membrane is called the "retentate" phase. Due to varied solubilities and diffusion rates of the components, distinct transport rates of the molecules are observed, which in turn facilitate the selectivity of the separation. Figure 4.3 depicts a schematic of the separation of a multi-component mixture using the pervaporation process. The location of the vapor-liquid interface can be anywhere within the membrane and sometimes even at a position very near to the permeate or feed side.

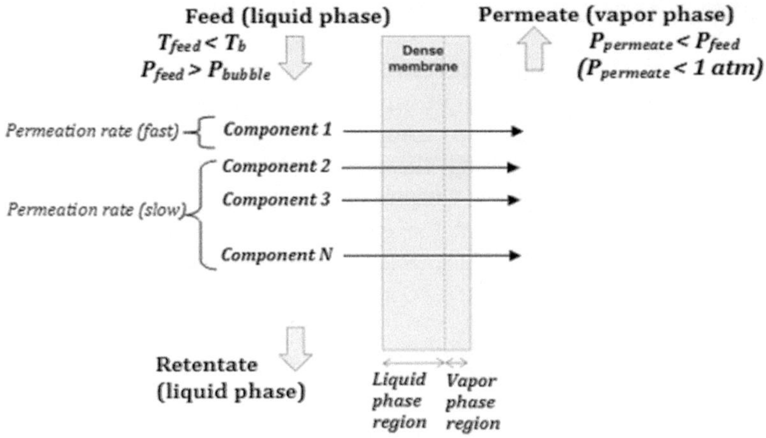

Fig. 4.3: Pervaporation process of a multicomponent mixture with N components. T_{feed} – feed temperature; T_b – boiling point of the feed solution; P_{feed} – feed pressure; P_{bubble} – bubble point of the feed solution; $P_{permeate}$ – permeate pressure

The evaluation of the mass transfer in the pervaporation process can be executed in two approaches. The first approach involves determining the transmembrane flux (individual flux of each component and total permeate flux) and the separation factor. Essentially, the transmembrane flux and the separation factor are determined from experimental data collected during membrane testing under specified operating conditions, including feed composition, pressure, and temperature. This will provide information about the area of the membrane required to achieve a specified separation and the desired purity of permeate. Thus, the process performance can be assessed. Conversely, the second approach involves adapting mathematical models to describe mass transfer through the membrane. This leads to the evaluation of the mass transfer coefficient (M.T.C.), permeability, and selectivity. Compared to the first approach, this allows for a deeper interpretation and evaluation of the extent to which the membrane increases the permeation of one component over another. It also provides insight into whether the membrane selection and the process operation are optimal. Therefore, this approach provides information on membrane separation performance. Both approaches provide complementary information, which is essential for understanding the overall scenario regarding separation efficiency and the optimum operating conditions that should be maintained during the pervaporation process. Based on critical reasoning and a preliminary overview of process applications, essential information must be gathered for performing pervaporation as a separation technology and for synthesizing specific membranes for a particular separation target. Proper understanding of the differences in the usage of parameters such as transmembrane flux, selectivity, separation factor, and permeability is of utmost importance for advancing pervaporation technology to the industrial level.

4.4 Mass Transport in Gas Separation Process

In the case of a gas separation process, the pressure (p_1) on the permeate side is comparatively lower than the pressure on the feed side (p_0). This creates a pressure gradient between the two compartments of the pervaporation module, facilitating gas separation through the membrane. To derive the gas transport expression, equation 2.4 (refer to Chapter 2) may be considered for compressible gases. The chemical potential at the gas side will be equal to the chemical potential at the membrane interface. Substituting this concept into equation 1.4, the equation becomes:

$$\mu_i^0 + RT\ln\left(\gamma_{i0} n_{i0}\right) + RT\ln\left(\frac{p_0}{p_i^{sat}}\right) = \mu_i^0 + RT\ln\left(\gamma_{i0(m)} n_{i0(m)}\right) + \sigma_i\left(p_0 - p_i^{sat}\right) \tag{4.1}$$

Rearranging the equation 4.1 yields,

$$n_{i0(m)} = \frac{\gamma_{i0}}{\gamma_{i0(m)}} . \left(\frac{p_0}{p_i^{sat}}\right) . n_{i0} . e^{\frac{-\sigma_i\left(p_0 - p_i^{sat}\right)}{RT}} \tag{4.2}$$

For gases, the value of the exponential term can be considered as approximately one even in the very high pressure cases. Therefore, equation 4.2 reduces to

$$n_{i0(m)} = \frac{\gamma_{i0}}{\gamma_{i0(m)}} . \left(\frac{p_0}{p_i^{sat}}\right) . n_{i0} \tag{4.3}$$

The term 'n_{i0} P_0' signifies the partial pressure (p_{i0}), of the 'i'th component of the feed stream (gas). So, equation 4.4 is simplified to

$$n_{i0(m)} = \frac{\gamma_{i0}}{\gamma_{i0(m)}} . \left(\frac{p_{i0}}{p_i^{sat}}\right) \tag{4.4}$$

This equation can be expressed in the terms of molar concentration as given below:

$$c_{i0(m)} = M_{wi} . \sigma_m . \frac{\gamma_{i0}}{\gamma_{i0(m)}} . \left(\frac{p_{i0}}{p_i^{sat}}\right) \tag{4.5}$$

The sorption coefficient for gas sorption (k_i) is expressed as

$$k_i = M_{wi} . \sigma_m . \frac{\gamma_{i0}}{\gamma_{i0(m)}} . \left(\frac{\delta_m}{p_i^{sat}}\right) \tag{4.6}$$

So, the concentration of the 'i'th component at the interface of feed–membrane, may be represented as

$$c_{i0(m)} = k_i . p_{i0} \tag{4.7}$$

Similarly, the concentration of the 'i'th component at the interface of membrane-permeate, may be represented as

$$c_{il(m)} = k_i . p_{il} \tag{4.8}$$

Combining the equations 4.7 and 4.8 and then substituting it in the Fick's law expression, the flux can be written as

$$J_i = \frac{D_i k_i \left(p_{i0} - p_{il}\right)}{l} \tag{4.9}$$

$$J_i = \frac{\alpha_i \left(p_{i0} - p_{il}\right)}{l} \tag{4.10}$$

where, α_i signifies the gas permeability coefficient which equals the term '$D_i k_i$'. This equation is majorly utilized for predicting the gas separation based membrane separation process.

4.5 Mass Transport in Pervaporation

In the pervaporation process, the feed is introduced into the module in the liquid phase. However, after permeation through the membrane, the permeant is available in the vapor phase. The driving force for this process is the partial pressures of the feed components. To maintain the significant driving force required to run the pervaporation process smoothly, either a vacuum or a temperature difference is generated in the permeate compartment, at laboratory and industrial scales respectively. Similar to the gas separation, the chemical potential at the feed side is assumed to be equal to that at the membrane interface at constant pressure for developing the expression of flux in the pervaporation process. Therefore, in this case equation 1.3 from chapter 1 will yield:

$$\mu_i^0 + RT \ln\left(\gamma_{i0} n_{i0}\right) + \sigma_i \left(p_0 - p_i^{sat}\right) = \mu_i^0 + RT \ln\left(\gamma_{i0(m)} n_{i0(m)}\right) + \sigma_i \left(p_0 - p_i^{sat}\right) \tag{4.11}$$

The concentration at the feed side of the interface is given by:

$$c_{i0(m)} = \frac{\sigma_{i0} \delta_m}{\sigma_{i0(m)} \delta_0} c_{i0} = K_i^L . c_{i0} \tag{4.12}$$

where, 'L' in superscript for denotes the liquid phase and K_i^L represents the sorption coefficient of the feed stream.

The pressure drops from p_0 at the membrane to p_l at the vapor phase. In such a case, the chemical potential of each phase can be expressed as follows:

$$\mu_i^0 + RT \ln\left(\gamma_{il} n_{il}\right) + RT \ln \frac{p_l}{p_i^{sat}} = \mu_i^0 + RT \ln\left(\gamma_{il(m)} n_{il(m)}\right) + \sigma_i \left(p_0 - p_i^{sat}\right) \tag{4.13}$$

Rearranging the equation gives,

$$n_{il(m)} = \frac{\gamma_{il}}{\gamma_{il(m)}} \cdot \left(\frac{p_l}{p_i^{sat}} \right) . n_{il} . e^{\frac{-\sigma_i \left(p_0 - p_i^{sat} \right)}{RT}} \tag{4.14}$$

Similar to the gas separation process, the behavior of this pervaporation system is the same. Hence the exponential term is neglected as its value is nearly one. Therefore, the concentration in the permeate phase can be represented by,

$$n_{il(m)} = \frac{\gamma_{il}}{\gamma_{il(m)}} \cdot \left(\frac{p_l}{p_i^{sat}} \right) . n_{il} \tag{4.15}$$

The term ' $n_{il}\ p_l$ ' represents the partial pressure (p_{il}) of the '*i*'th component. So, Equation 4.15 simplifies to,

$$n_{il(m)} = \frac{\gamma_{il}}{\gamma_{il(m)}} \cdot \left(\frac{p_{il}}{p_i^{sat}} \right) \tag{4.16}$$

Now, in terms of concentration, the equation can be expressed as,

$$c_i^L = M_{wi} . \delta . \frac{\gamma_i^G}{\gamma_i^L} \cdot \left(\frac{p_i}{p_i^{sat}} \right) = \frac{K_i^G}{K_i^L} \cdot p_i \tag{4.17}$$

This equation can also be expressed in the terms of molar concentration as,

$$c_{il(m)} = M_{wi} . \delta_m \cdot \frac{\gamma_{il}}{\gamma_{il(m)}} \cdot \left(\frac{p_{il}}{p_i^{sat}} \right) \tag{4.18}$$

The sorption coefficient for gas sorption (K_i^G) is expressed as

$$K_i^G = M_{wi} . \delta_m \cdot \frac{\gamma_{il}}{\gamma_{il(m)}} \cdot \left(\frac{1}{p_i^{sat}} \right) \tag{4.19}$$

The equation 4.18 can be reduced by substituting the equation 4.19 to get,

$$c_{il(m)} = K_i^G . p_{il} \tag{4.20}$$

where K_i^G signifies the sorption coefficient of gas phase and G is used in superscript for denoting the gas phase.

Now, for deriving the final equation for the permeate flux in the pervaporation process, the substitution of equations 4.12 and 4.20 into the Fick's law expression gives,

$$J_i = \frac{D_i \left(K_i^L c_{i0} - K_i^G p_{il} \right)}{l} \tag{4.21}$$

However, two different sorption coefficients (gas and liquid phase coefficients) are present in the above equation. Henceforth, it is essential to develop an expression which handles both separation coefficients without difficulty. In such a scenario it is assumed that hypothetical vapor is in equilibrium with the liquid feed stream. This VLE will provide the following relationship:

$$\mu_i^0 + RT\ln\left(\gamma_i^L n_i^L\right) + \sigma_i\left(p - p_i^{sat}\right) = \mu_i^0 + RT\ln\left(\gamma_i^G n_i^G\right) + RT\ln\frac{p}{p_i^{sat}} \tag{4.22}$$

The denotation of liquid and vapor phases is done by the superscripts '*L*' and '*G*'. Following the same procedure from equations 4.13 to 4.20, the equation becomes

$$n_i^L = \frac{\gamma_i^G}{\gamma_i^L}\cdot\left(\frac{p_i}{p_i^{sat}}\right) \tag{4.23}$$

The expression can be expressed in terms of concentration as follows:

$$c_i^L = M_{wi}.\delta.\frac{\gamma_i^G}{\gamma_i^L}\cdot\left(\frac{p_i}{p_i^{sat}}\right) = \frac{K_i^G}{K_i^L}\cdot p_i \tag{4.24}$$

This expression relates the concentration of the '*i*'th component in the feed at the liquid phase (c_i^L) to the partial vapor pressure (p_i) of the '*i*'th component which is in equilibrium with the liquid phase. Now, the flux equation can be written as

$$J_i = \frac{D_i K_i^G\left(p_{i0} - p_{il}\right)}{l} \tag{4.25}$$

where p_{i0} and p_{il} denote the partial vapor pressures of the '*i*'th component present on two different membrane sides. Equation 4.25 may be rewritten as

$$J_i = \frac{P_i^G}{l}\left(p_{i0} - p_{il}\right) \tag{4.26}$$

Equation 4.26 can be utilized for evaluating the vapor pressure difference which acts as the driving force for the pervaporation membrane process.

4.6 Why is the Pervaporation Process Preferred?

As discussed in section 4.1, pervaporation is a three-step separation process that includes sorption, diffusion, and desorption. This process is compatible even with azeotropic (same boiling-range) mixtures and has the ability to mitigate the shortcomings of VLE. Furthermore, the applicability of pervaporation gets enhanced when combined with another separation technology (like distillation, or adsorption) in a hybrid process. The major benefits of this process are listed below:

- In normal distillation, separation of azeotropes demands a large number of theoretical stages. However, in pervaporation, the collection of specified components of a mixture in the permeate stream is mainly based on the membrane attributes. Separation of such azeotropic mixtures is possible without using any entrainers. Since an entrainer is not used, high product purity at a comparatively lower capital investment is achievable along with the added advantage of no environmental pollution risk.

- Effective dehydration of a multi-component aqueous mixture having nearly equal boiling points.
- Treatment of the feed mixtures may be performed in either liquid or vapor phase.
- Less energy requirement for both the vapor and liquid form.
- When pervaporation is performed in hybrid mode, in conjunction with other technologies like distillation, energy consumption is significantly reduced. Moreover, modular design also lowers operational expense.
- It acts as a multipurpose and highly versatile system in context with the accommodation of feed mixtures followed by the production of high purity products.
- Compact design, minimal area requirement, easy and time-effective installation, uncomplicated operation, and simple startup and shutdown procedures.

Considering these advantages, the pervaporation process is mainly used in the chemical process industries. It also finds application in areas such as food industry, pharmaceutical industry, environmental treatment, and analytical applications. The preference for pervaporation technology lies in its ability to treat specific substances that present complexities when using other conventional methods. Some common applications are presented in Table 4.1.

Table 4.1: Some common applications of the pervaporation separation process adopted in various industries

Aqueous mixture	Non-aqueous mixture
Dehydration (removal of water from organic solvents)	Alcohols/aromatics (methanol/toluene)
Separation of alcohols (ethanol, butanol) from fermentation broths	Alcohols/aliphatics (ethanol/hexane)
Separation of volatile organic contaminants (aromatic compounds, chlorinated hydrocarbons) from wastewater	Alcohols/ethers (Methanol/MTBE)
Separation of phenolic compounds	C-8 isomers (o-xylene, m-xylene, p-xylene, styrene)
Separation of aroma and flavor compounds	Aromatics/Aliphatics (e.g. cyclohexane/benzene)
	Saturated/Unsaturated (e.g. butane/butane)

4.7 Process Design Parameters of Pervaporation Process

The design parameters for a pervaporation setup at the laboratory scale often differ from those at the industrial scale, as they become technically infeasible or unrealistic for industrial applications. Ideal fluid dynamic conditions are commonly considered in lab-scale setups. However, to ensure the transport of target components at the industrial level, optimum driving force conditions may require cost-intensive high cross-flow velocities or dimensioned vacuum pumps. Thus, in-depth fundamental knowledge and proper selection of operating conditions are prerequisite for designing an effective process.

4.7.1 Process Parameters of the Feed Side

In some cases, due to the unresponsive nature of the liquid mixture, the driving agent is unable to ensure transport even when rigorous pressure is applied to the liquid stream. This is because a substantial increase in pressure does not significantly alter concentration, unlike in vapor permeation. Hence, the pervaporation process is limited by a low driving force due to the concentration of the feed component intended to permeate across the pervaporation membrane module, which eventually reduces the flux. In such situations, the foremost requirement is the assessment of optimal parametric conditions favorable for the mass transfer of the component of interest at the interface of the feed and membrane. In this regard, proper membrane selection demands higher affinity towards the membrane material, adequate partitioning, and sufficient driving force to ensure diffusion of the target components across the membrane. If the components of interest have high partitioning capability towards the membrane in comparison to the bulk solvent, these components may accumulate at the feed/membrane interface resulting in hindrance of the consequent separation procedure. The elevation of the components' concentration at the interface may be restrained by reducing the thickness of the mass transfer boundary layer (MTBL). As we discussed in chapter 1, in the pressure-driven membrane process, the permeation of solvent mostly takes place only with meagre amounts of solutes. However, in the pervaporation process, the component of interest exhibits greater partitioning to the membrane than the solvent. Hence, during the pervaporation process, the reduction of the component of interest adjacent to the membrane surface signifies the depletion in local concentration at the interface of feed and membrane. This eventually creates a negative impact on the driving force for transportation of solute.

The effect of the mass transfer layer can be reduced by proper system design and maintaining optimal parametric conditions related to fluid dynamics. However, implementing favorable fluid dynamic conditions mandates higher energy investment in the pervaporation system. Previous research has shown that using cross-flow at elevated velocities can mitigate external mass transfer limitations for components with less partitioning. Conversely, the same cross-flow conditions were not found to be effective in completely mitigating external mass transfer hindrances for components with substantial partitioning to the membrane. This indicates that the derived permeate flux remains inadequate even after implementing very high cross-flow velocities. From an industrial perspective, the application of cross-flow velocities is desired to achieve very high permeation flux with minimal energy input. Moreover, specially designed spacers are used to generate turbulence at the membrane surface. The energy input required for these spacers must be taken into account. Furthermore, a critical analysis must be performed to evaluate the flux increase relative to a definite energy investment.

The use of the thermo-pervaporation process can be a viable option for providing significant driving force to execute the transportation of desired solutes. By preheating the feed stream, the partial pressure of different components in the feed stream is enhanced, thereby developing a gradient by keeping the permeate side at a much lower temperature than the feed compartment. This eventually promotes the driving force required for transportation of molecules. This process is already discussed elaborately in section 4.2.

4.7.2 Process Parameters of the Downstream Compartment

In the downstream compartment, either a sweeping gas or vacuum condition is used to provide the driving force for solutes to pass through the membrane. This lowers the vapor pressure of the permeated components, creating a partial pressure gradient that removes compounds across the membrane. When the partial pressure of the component to be transported is low, vacuum pervaporation is preferred. Conversely, sweeping gas pervaporation requires a very high flow rate of sweep gas to ensure the permeation of solutes with low partial pressure. This method is particularly effective when the partial pressure of the component is 20 mbar or higher.

It's crucial to prevent condensation of components with low permeation capability when passing through the porous support of a pervaporation membrane. Inefficient transport can lead to local elevation of partial pressures, known as concentration polarization, typically observed on the downstream side of the membrane. To control this issue, a high carrier gas flow rate is needed for sweeping gas pervaporation, while a low intensity vacuum condition is required for vacuum pervaporation.

A comparative study evaluated sweeping gas and vacuum pervaporation using a dense membrane (82). Sweeping gas operation consumes less energy, making it more economically feasible, is consumed by sweeping gas operation than by vacuum pervaporation. However, despite its high energy requirements, vacuum pervaporation is preferred for separating and recovering pure compounds. Due to its high cost, employing vacuum evaporation at extremely low intensities is an infeasible strategy. The vacuum pressure in the downstream permeate compartment of pervaporation module is typically maintained between 10–80 mbar, but this may vary based on the permeation characteristics.

Consider a scenario where the pervaporation process operates alongside a fermenter containing significant amounts of dissolved non-condensable gases like hydrogen and carbon dioxide. Before starting, a vacuum pump should remove these gases. Following this, sweeping gas pervaporation becomes a favorable option.

4.8 Membranes for Pervaporation

Pervaporation membranes are categorized into three main types: hydrophobic, hydrophilic, and organophilic. Hydrophilic membranes are commonly used to separate water from organic solvents such as polyacrylic acid (PAA), polyvinyl alcohol (PVA), polyacrylonitrile (PAN), and polyvinylpyrrolidone (PVP). They are effective in separating water present in dilute concentrations within azeotropic mixtures. Hydrophilic membranes are used for dehydrating multicomponent and azeotropic mixtures like water/ethanol, water/isopropanol, and dimethylformamide (DMF) with less than 10% water. Hydrophobic membranes, such as polytrimethylsilyl-1-propyne (PTMSP), polydimethylsiloxane (PDMS), and polyvinylidenefluoride (PVDF), remove low-concentration volatile organic compounds from aqueous solutions. Their applications include recovering solvents in the pharmaceutical industry, extracting aroma compounds in the food industry, removing trihalomethanes (THM) and methyl-tert-butyl ether (MTBE) from groundwater, and treating industrial wastewater. Organophilic membranes extract organic compounds from aqueous solutions and separate multicomponent or binary organic mixtures, including polar-nonpolar, alicyclic-aromatic, and isomeric compounds. These organophilic membranes are utilized to separate a mixture of aldehydes, amines, and alcohols, and ethanol from benzene, etc.

4.9 Industrial Scenario and Future Prospects

The global market for pervaporation membranes grew from over 10 million USD in 2000 to approximately 1.1 Billion USD in 2021 and is expected to grow at a CAGR of 9.6% from 2021 to 2023. Commercially available pervaporation modules include plate frames, spiral modules, and inorganic multichannel tubes. Leading companies in the industry are Sulzer Chemtech GmbH, Compact Membrane Systems, Vladipor Ltd., Pervatech, Mitsui Zosen Machinery & Service Inc., BUSS-SMS-Canzler, Ube Industries Ltd., and the Energy research Centre of the Netherlands (ECN). Sulzer Chemtech GmbH alone had installed over 200 pervaporation systems by 2010, primarily for dehydrating alcohols and solvents in chemical and pharmaceutical industries. Ongoing research focuses on developing novel membranes and improving membrane attributes to enhance the efficiency of industrial liquid separations. Separating organic mixtures via pervaporation is challenging due to the similar properties of solutes and the harsh operating conditions. In the petrochemical industry, removing sulfur and aromatic components from fuels is crucial. Research on suitable membranes, particularly mixed-matrix membranes is essential to ensure they can handle severe conditions during prolonged exposure. To separate organic compounds present at very low concentrations, organophilic pervaporation must be developed. The system should be designed to create a generally low driving force due to the lower concentration of target compounds in the feed solution. Developing membrane modules with a lower pressure drop on the feed side is essential to counteract external mass transfer hindrances. Additionally, integrated separation technology is crucial for efficiently recovering valuable components like natural aroma from aqueous solution or fermentation media. Hybrid technologies incorporating absorption, fractionated condensation, distillation, encapsulation within nanostructured materials, and adsorption alongside pervaporation hold promise. In conclusion, pervaporation is a promising membrane separation process requiring advancements in membranes to improve permeation flux, selectivity, long-term stability, and cost-effectiveness.

4.10 Summary

- A brief on gas separation and pervaporation.
- Elaboration on different types of pervaporation.
- Mass transport model for gas separation and pervaporation.
- Discussion on the membranes used for pervaporation and industrial scenario for pervaporation.

CHAPTER

5

Liquid Membrane Separation

5.1 Understanding Liquid Membranes and their Significance

A liquid membrane is a thin film that separates two phases to facilitate the separation of desired or undesired components from a liquid, often for pollutant extraction from wastewater. One of the fundamental principles of liquid membranes is creating an interface between two immiscible liquids. There are two basic types of liquid membranes: emulsion liquid membranes and supported liquid membranes. Emulsion liquid membranes are less stable, whereas supported liquid membranes consist of a solid membrane with pores filled by the extracting phase. Why choose a liquid membrane over a conventional one? Consider the example of separating metal ions, like chromium, from wastewater. Conventional membranes, such as RO membranes, are costly. Liquid membranes offer a more cost-effective alternative for such separations.

Opting for the micellar enhanced ultrafiltration (MEUF) process, as discussed in Chapter 7, involves complexities due to two main rate-limiting stages: the entrapment percentage of pollutants within the micelle and the separation of micellized pollutants during ultrafiltration. In contrast the removal percentage with either type of liquid membrane depends solely depends on a single stage operation – the transfer across the interface. Detailed discussions on different types of liquid membranes, their fabrication technology and applications are provided in the subsequent sections.

5.2 Emulsion Liquid Membrane (ELM)

Emulsion liquid membrane (ELM) involves mixing an immiscible organic phase (lean) with an aqueous phase (rich) using a surfactant. Surfactants have a polar (hydrophilic) end and an apolar (lipophilic) end, and their purpose of the surfactant is to reduce surface tension when added to a liquid. Understanding the preparation of ELM requires knowledge of basic surfactant chemistry and emulsions. An emulsion is a mixture of two immiscible liquids, typically an organic liquid and water. There are different types: water-in-oil (W/O), oil-in-water (O/W) and water-in-oil-in-water (W/O/W). The type of emulsion formed is controlled by the surfactant's HLB (hydrophilic-lipophilic balance) value. HLB indicates the hydrophilic contribution in a surfactant molecule. A higher HLB means greater contribution, leading to more water molecules forming a continuous aqueous phase with a discrete organic phase, creating an O/W emulsion. Conversely, a lipophilic contribution forms an organic phase with discrete aqueous phase, resulting in a W/O emulsion (Figure 5.1). In the case of ELM, the extractant is the organic phase where pollutants accumulate from the continuous wastewater phase and then transfer to an internal solvent (stripping) phase that solubilizes the pollutants. The middle organic phase acts as the membrane, facilitating the transfer of pollutants from the external feed wastewater to the stripping phase (Figure 5.2).

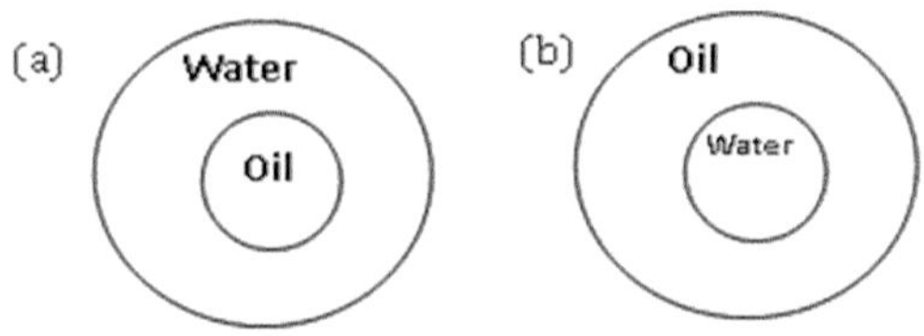

Fig. 5.1: (a) W/O emulsion, (b) O/W emulsion

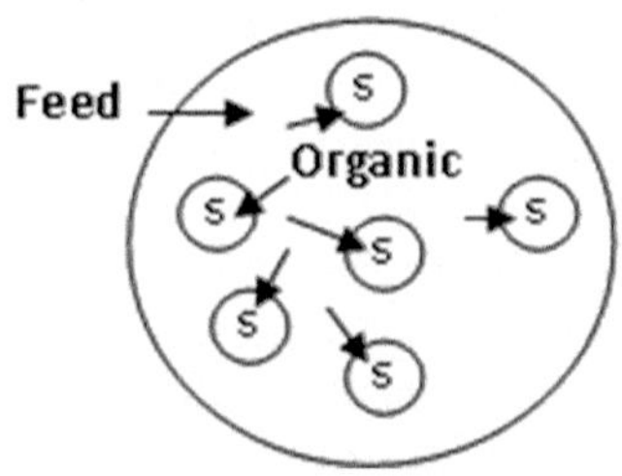

Fig. 5.2: Mass transport in ELM

5.2.1 Preparation of ELM and Transport Mechanism

Let's consider a case study to understand the basic requirements for preparing an ELM. Three components are needed to create an effective liquid membrane: diluents, an organic agent or carrier (membrane phase), a surfactant and a stripping agent. The primary strategy for preparing an ELM involvesformulating an emulsion by mixing diluents, the carrier and a surfactant with the stripping phase. The HLB value of the surfactant should be in the range of 4–5 to ensure a W/O emulsion. During the extraction process, this prepared ELM is introduced into the feed containing pollutants, which are extracted into the stripping phase with appropriate stirring. Ahmad et al. (84) elaborated a process to extract acetaminophen, an antibiotic found in pharmaceutical wastewater. They used trioctylamine (TOA) as the carrier, Span 80 as the surfactant and kerosene, *n*-heptane and *o*-xylene as the diluents, and nitric acid, ammonia, and acetic acid as the stripping agents. All components were mixed at 500 rpm, followed by homogenization using ultrasonication. To extract acetaminophen (ACTP) from wastewater, the formulated ELM was introduced into the pollutant-rich wastewater and stirred to enhance the extraction rate. In this case, the transport from the external membrane interface to the internal membrane interface is facilitated by the carrier, moving the pollutant into the organic phase. In this case, ACTP combines with TOA to form a complex that is transported into the stripping phase. One of the primary concerns with this transport is the nature of the diluents used for the diffusion of the species from the external to the internal interface of the ELM. The highest extraction efficiency, nearly 82%, was observed with kerosene. Similarly, at the internal interface (interface between the stripping agent and organic liquid) the efficiency of the stripping agent is crucial for extracting pollutants from the complex and releasing the carrier. Ammonia achieved 82% stripping efficiency, compared to 57% with nitric acid and 73% with acetic acid. Figure 5.3 shows a schematic for the facilitated transport of the pollutants. Mohammed et al. (85) had utilized an organic surfactant egg yolk to form the emulsion membrane to extract phenol from the aqueous solution. They used xylene as the diluent, with egg yolk dissolved in it, and caustic as the stripping phase. 82% of the phenol was removed with a stripping efficiency of 91.1%. A critical aspect of any ELM is the breaking of the emulsion and infusion of the internal phase into the feed side. The breaking percentage is calculated as the ratio of the volume of the internal phase leaked into the feed phase to the initial volume of the internal phase. The leaked volume phase is calculated by equation 5.1 that follows.

$$V_{\text{internal}} = \frac{V_{\text{feed phase initial}}\left(10^{-\text{pH}_0} - 10^{-\text{pH}(t)}\right)}{\left(10^{-\text{pH}(t)} - C_{\text{OH}}^{\text{internal}}\right)} \tag{5.1}$$

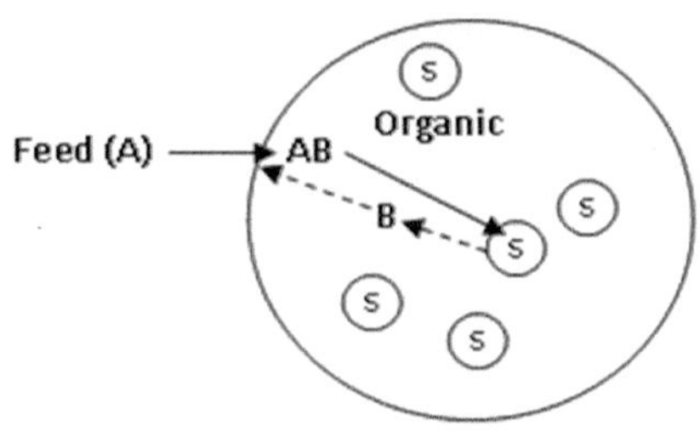

Fig. 5.3: Facilitated transport of pollutant A through a mediator B inside the organic phase

Huang et al. (86) derived a basic model on ELM based on some idealizations with the following assumptions:

- Emulsion globules are monodisperse and spherical.
- Mass transfer within the membrane phase is only through diffusion and the effective diffusivity is constant throughout the membrane phase.

- There is no coalescence of emulsion globules or other phases.
- Internal mass transfer resistance is negligible.
- Negligible leakage of the internal phase into the external phase.

Based on this a mass balance has been made for the external and membrane phase.

$$V_e \frac{dC_e}{dt} = -N\left(4\pi R^2\right)k\left(C_e - C_m^*\right) = -N\left(4\pi R^2\right)D_e\left(\frac{\partial C_m}{\partial r}\right)\bigg|_{r=R} \tag{5.2}$$

$$C_m^* = K_D C_e \tag{5.3}$$

Within the membrane phase the mass balance equation is,

$$V_m \frac{\partial C_m}{\partial t} = \left(V_m + V_i\right)D_e\left[\frac{1}{r^2}\frac{\partial}{\partial r}\left(r^2\frac{\partial C_m}{\partial r}\right)\right] - V_i R_x \tag{5.4}$$

$(V_m + V_i)$ is the total volume of the emulsion, R_x is the rate of disappearance of the solute-carrier complex in the membrane phase into the stripping phase

$$= \frac{\partial C_i\left(r,t\right)}{\partial t}. \tag{5.5}$$

C_i is the concentration of the solute inside the stripping phase = qC_m. Combining equations 5.4 and 5.5, we get the following equation

$$\left(\frac{V_m + qV_i}{V_m + V_i}\right)\frac{\partial C_m}{\partial t} = D_e\left[\frac{1}{r^2}\frac{\partial}{\partial r}\left(r^2\frac{\partial C_m}{\partial r}\right)\right] \tag{5.6}$$

Before approaching the solution to equation 5.6, let's first conceptualize some of the terms described here. First, what is Sauter mean radius? It is the mean diameter calculated by averaging the individual colonies of bubbles or particles of the same diameter (87) as given in equation 5.7.

$$d_j = d_{32} = \frac{\sum_{i=1}^{n} n_i d_i^3}{\sum_{i=1}^{n} n_i d_i^2} \tag{5.7}$$

Now in order to solve the equation 5.6, the boundary conditions are,

$$t = 0; C_m = 0$$

$$t > 0, r = R_{s,m}; \frac{\partial C_m}{\partial r} = \frac{k}{D_e}\left(C_e - C_m^*\right) = \frac{kC_e}{D_e}\left(1 - K_D\right)$$

$$t > 0, r = R_{s,s}; C_m = {C_i}/{q}$$

where, $R_{s,m}$ = Sauter mean radius for membrane phase, $R_{s,s}$ = Sauter mean radius for stripping phase.

Upon further simplification, it can be assumed that the value of q is much higher in this case, indicating maximum accumulation of pollutants within the stripping phase. Since ‘r’ and ‘t’ are independent of each other, the separation of variables technique can be applied to obtain an analytical solution. Accordingly, one can write, $C_m = C_m(r, t) = A(r)B(t)$.

On applying the technique in equation 5.6,

$$QAB' = D_e\left[\frac{1}{r^2}\frac{\partial}{\partial r}\left(r^2\frac{\partial C_m}{\partial r}\right)\right] = D_e\left[\frac{\partial^2 C_m}{\partial r^2} + \frac{2}{r}\frac{\partial C_m}{\partial r}\right] = D_e B\left(A'' + \frac{2}{r}A'\right)$$

where, $Q = \dfrac{V_m + qV_i}{V_m + V_i}$

$$\therefore \frac{QB'}{D_e B} = \frac{A''}{A} + \frac{2}{r}\frac{A'}{A} = -\frac{\lambda^2}{r^2} \tag{5.8}$$

Therefore after formulating the equation it is found that,

$$\left.\begin{aligned} B' + \left(\frac{\lambda^2 D_e}{r^2 Q}\right)B = 0 \\ A' + \frac{2}{r}A' + \frac{\lambda^2}{r^2}A = 0 \end{aligned}\right\} \tag{5.9}$$

Now,

$$\frac{dB}{dt} = \left(-\frac{\lambda^2 D_e}{r^2 Q}\right)B \tag{5.10}$$

$$\therefore\ B = \exp\left(-\frac{\lambda^2 D_e}{r^2 Q}\right)t + C_1$$

$$\text{At } t = 0,\ B = 0\ \therefore C_1 = -1 \Rightarrow B = \exp\left(-\frac{\lambda^2 D_e}{r^2 Q}\right)t - 1$$

Further,

$$A'' + \frac{2}{r}A' + \frac{\lambda^2}{r^2}A = 0 \Rightarrow r^2 A'' + 2rA' + \lambda^2 A = 0 \tag{5.11}$$

Equation 5.11 is solved using the Cauchy-Euler substitution technique. The general solution is $A = -\dfrac{1}{2} \pm \dfrac{\sqrt{1-4\lambda^2}}{2}$. Now to comment on the nature of the roots one must draw some idea on the value of ‘L’, which is an eigenvalue. What is an eigenvalue? It essentially describes the fate of the output after traveling through space or time. In this context, the eigenvalue varies with space and must be assigned to reflect the decrease in concentration, maintaining a proper gradient from the membrane to the stripping phase. This value is significant for understanding the substantial effect on the outcome. Therefore the roots must be complex and the solution is given by,

$$A = r^{-\frac{1}{2}}\left[C_1 \cos\left(\frac{\sqrt{1-4\lambda^2}}{2}\ln r\right) + C_2 \sin\left(\frac{\sqrt{1-4\lambda^2}}{2}\ln r\right)\right] \tag{5.12}$$

Now applying boundary conditions for '*r*',

$$A = r^{-\frac{1}{2}}\left[\frac{M}{\cos\left(\frac{\sqrt{1-4\lambda^2}}{2}\ln R_{s,m}\right) - \sin\left(\frac{\sqrt{1-4\lambda^2}}{2}\ln R_{s,m}\right)\cos\left(\frac{\sqrt{1-4\lambda^2}}{2}\ln R_{s,s}\right)}\cos\left(\frac{\sqrt{1-4\lambda^2}}{2}\ln r\right) + \frac{M}{\sin\left(\frac{\sqrt{1-4\lambda^2}}{2}\ln R_{s,m}\right) - \cos\left(\frac{\sqrt{1-4\lambda^2}}{2}\ln R_{s,m}\right)\sin\left(\frac{\sqrt{1-4\lambda^2}}{2}\ln R_{s,s}\right)}\sin\left(\frac{\sqrt{1-4\lambda^2}}{2}\ln r\right)\right]$$

$$M = -\frac{2kC_e R_{s,m}^{1.5}}{D_e}\left(1-K_D\right) \tag{5.13}$$

Hence,

$$C_m(r,t) = r^{-\frac{1}{2}}\sum_n \left[\exp\left(-\frac{\lambda_n^2 D_e}{r^2 Q}\right)t - 1\right]\left[\frac{M}{\cos\left(\frac{\sqrt{1-4\lambda_n^2}}{2}\ln R_{s,m}\right) - \sin\left(\frac{\sqrt{1-4\lambda_n^2}}{2}\ln R_{s,m}\right)\cos\left(\frac{\sqrt{1-4\lambda_n^2}}{2}\ln R_{s,s}\right)}\cos\left(\frac{\sqrt{1-4\lambda_n^2}}{2}\ln r\right) + \frac{M}{\sin\left(\frac{\sqrt{1-4\lambda_n^2}}{2}\ln R_{s,m}\right) - \cos\left(\frac{\sqrt{1-4\lambda_n^2}}{2}\ln R_{s,m}\right)\sin\left(\frac{\sqrt{1-4\lambda_n^2}}{2}\ln R_{s,s}\right)}\sin\left(\frac{\sqrt{1-4\lambda_n^2}}{2}\ln r\right)\right] \tag{5.14}$$

5.3 Supported Liquid Membrane (SLM)

The basic difference between ELM and SLM is the presence of a conventional membrane. In SLM, the organic phase is inserted into the pores of the membrane during the immersion of polymeric membrane in organic solvent, acting as a separation barrier for the pollutants. On the feed side, the pollutant rich phase passes through, and pollutants are captured by the organic phase due to the concentration gradient created at the interface of the feed and the organic liquid. On the permeate side of the membrane, the phase, called the stripping phase, collects pollutants from the organic phase. Continuous removal of pollutants with the stripping phase from the membrane module maintains a constant gradient across the interface. This membrane is also called an immobilized liquid membrane (ILM) due to the immobilization of liquid within the membrane. One significant application of SLM is the extraction of metal ions through the organic phase embedded within the membrane pores. The basic principle in separating metal ions through the embedded organic phase within the pores is carrier facilitated transport. A carrier attaches to the pollutant and aids its movement towards the stripping phase. The transport mechanism can be categorized into two segments: cotransport and countertransport (88). In countertransport, metals attach to the carrier present in the organic solvent at the interface between the feed and the organic solvent. The metal is then transported toward the interface between stripping phase and the solvent, where it detaches from the carrier and moves into the stripping phase. In co-current transport, the carrier and pollutants are present in the feed and dissolve simultaneously into the organic phase. The assembled structure is carried from the feed side, to the stripping side followed by the release of both the pollutant and carrier into the stripping phase. Figure 5.4 describes the two processes for the same.

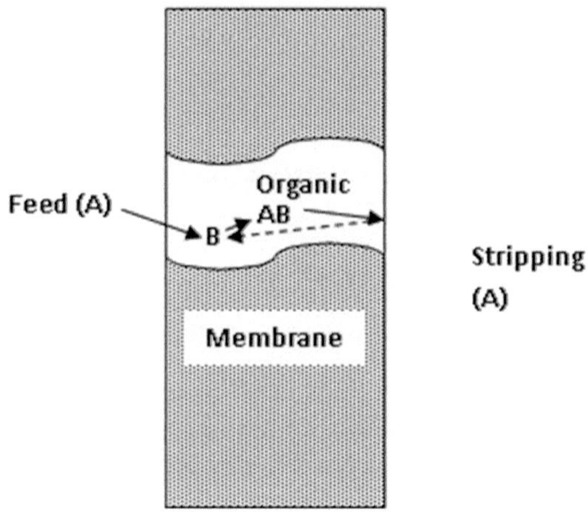

Fig. 5.4: Facilitated transport with SLM

Let's first discuss some carriers used for transporting pollutants across the membrane. Phenolic oxime extractant, a chelating compound, entraps metal from water and facilitates its transportation to the other side of the membrane. Figure 5.5 shows the entrapment of metal ions in it. This happens with the co-current transport of metal ions. In the countercurrent transport mechanism, strong ligand type counterions can extract metal after formation of the salt and thus solubilize the metal ions into the carriers after replacing the water molecules surrounding to make it more lipophilic. The mechanism of solubilization occurs when organophosphorus (phosphorus-oxygen bonding) group carriers are used. Conversely, carriers with carbon-oxygen bonding bind the metal solute by utilizing water molecule to form hydrogen bonds between the carrier and metal ions. This creates hydrated metal complexes in the organic membrane phase.

However, SLM is not only limited to the extraction of the metal ions, but also used to pharmaceuticals from aqueous solutions (89), and rare earth elements from soil (90). Drugs that are mostly found in wastewater are anti-inflammatory, antibiotics, and anticonvulsants. The most common drugs discussed for removal using a supported liquid membrane from wastewater are diclofenac, ibuprofen and carbamazepine. Farah et al. (89) used kerosene as the extractant with different additives like Cyanex 923 (Cy923, 94%, Solvay), trioctylamine (TOA, 98%), tributyl phosphate (TBP, 99%, Merk) and neodecanoic acid (Versatic Acid 10, 98 %). It was seen that 20% TBP and 10% Cy923 suitably extracted substantial amounts of diclofenac and ibuprofen at the basic pH. Conversely, the extraction of carbamazepine is low in all cases. Moreover, the separation will be performed based on a case-by-case formulation of the extractant in the case of SLM.

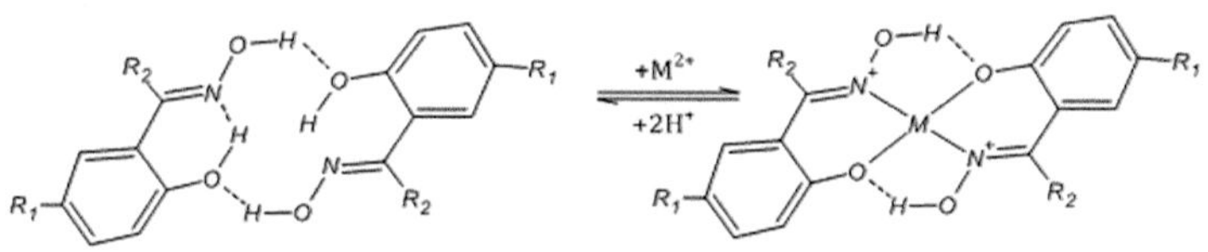

Fig. 5.5: Chelation of phenolic oxime extractant

Table 5.1 shows some common carriers used for the extraction of metals from wastewater.

5.3.1 Transport Mechanism in SLM

In general mass transport in SLM follows the following steps.

- Diffusion of metal ions occurs from the feed at the feed-SLM interface.
- Transportation by carrier at the SLM interface occurs through the formation of metal ion-carrier complex. For acidic carriers, hydrogen ions are released into the feed side; for basic or neutral carriers, anions from the feed side are co-extracted.
- Diffusion of metal ion-carrier complex within the SLM phase towards SLM-stripping side interface.
- Stripping through dissociation of metal ion-carrier complex releasing metal ions at the SLM-stripping solution interface (hydrogen ions replace the metal ions from the acidic carriers from stripping towards SLM).
- Diffusion of metal ion at SLM-stripping solution interface.

Lets accumulate the reaction briefly to precisely represent the above statements (91).

$$
\begin{aligned}
& M^{2+}(aq) + 2\,HR(org) \xrightleftharpoons{f-m,\ \text{interface}} R_2M(org) \\
& + 2H^+(aq) \xrightleftharpoons{\text{diffusion}} 2H^+(aq) + R_2M(org) \\
& \xrightleftharpoons{m-s,\ \text{interface}} M^{2+}(aq) + 2HR(org)
\end{aligned}
\tag{5.15}
$$

Table 5.1: Common carriers used for metal ions extraction (88)

Carrier	Carrier trade name	Carrier type	Extracted metal	Anion	Percentage extraction
Bis-(2,4,4-trimethylpentyl)-phosphinic acid	Cyanex 272	Acidic	Zn	SO_4^{2-}	94.39
			Co	SO_4^{2-}	99.99
Neodecanoic acid	Versatic 10	Acidic	Ag	SO_4^{2-}	85
			Cu	SO_4^{2-}	>90
Tri-n-butyl phosphate (TBP)	-	Basic/Solvating	Cr(VI)	Cl^-	99
2-hydroxy-5-nonylacetophenone oxime	LIX 841	Chelating	Mo(VI)	SO_4^{2-}	97.5
			Cu	SO_4^{2-}	~70

In one dimensional flow, the total mass transfer of metal on the feed and stripping sides is expressed as,

$$N_M = -D_M \frac{dC_M}{dz} + \frac{C_M}{C}\left(N_M + N_{H_2}\right) \tag{5.16}$$

For dilute solution, $C >> C_M$ and thus the convective term can be neglected. Within the membrane phase the total molar flux can be written as,

$$N_{M,m} = -D_{m,eff} \frac{dC_{M,m}}{dz} + \frac{C_{M,m}}{C}\left(N_{M,m} + N_{HR}\right) \tag{5.17}$$

As the carrier concentration is not varying with time, the second term can be neglected. Again the effective diffusivity ($D_{m,eff}$) through the membrane can be written as,

$D_{m,eff} = \frac{D_{R_2M}\varepsilon}{\zeta}$, where ζ is the tortuosity of the support and ε is the porosity of the support. Therefore, equations 5.15 and 5.17 can be written as, $N_M = -D_M \frac{dC_M}{dz}$ and $N_{M,m} = -D_{m,eff} \frac{dC_{M,m}}{dz}$ respectively, which in unsteady state yields,

$$\frac{\partial C_M}{\partial t} = D_M \frac{\partial^2 C_M}{\partial z^2} \tag{5.18}$$

$$\frac{\partial C_{R_2M}}{\partial t} = D_{m,eff} \frac{\partial^2 C_{R_2M}}{\partial z^2} \tag{5.19}$$

Now consider, $\theta = \frac{C(z,t)}{C_0}; \xi = \frac{z}{\partial}; \tau = \frac{Dt}{\partial^2}$, where ™ represents the thickness of the boundary layer. Now equations 5.18 and 5.19 can be re-written as,

$$\frac{\partial \theta_M}{\partial \tau} = \frac{\partial^2 \theta_M}{\partial \xi^2} \tag{5.20}$$

$$\frac{\partial \theta_{M,m}}{\partial \tau} = \frac{\partial^2 \theta_{M,m}}{\partial \xi^2} \tag{5.21}$$

Now let us elaborate the equation 5.20 as $\frac{\partial \theta_{M,f}}{\partial \tau} = \frac{\partial^2 \theta_{M,f}}{\partial \xi^2}$ for the feed and $\frac{\partial \theta_{M,s}}{\partial \tau} = \frac{\partial^2 \theta_{M,s}}{\partial \xi^2}$ stripping sides respectively. Now the boundary conditions for both sides can be expressed as,

Feed side:

$$\begin{cases} \theta_{M,f} = 0, \tau = 0 \\ \text{feed} - \text{membrane interface, feed side} \\ \theta_{M,f} = 0, \tau > 0, \xi = 1 \\ \text{at the top of the boundary layer due to reaction} \\ \theta_{M,f} = \exp\left(-\dfrac{k_1 \delta_f^2 \tau}{D_{M,f}}\right), \tau > 0, \xi = 0 \end{cases}$$

Stripping side:

$$\begin{cases} \theta_{M,s} = 0, \tau = 0 \\ \text{membrane stripping phase interface, stripping phase side} \\ \theta_{M,s} = \dfrac{k_1}{k_2 - k_1}\left[\exp\left(-\dfrac{k_1 \delta_s^2 \tau}{D_{M,s}}\right) - \exp\left(-\dfrac{k_2 \delta_s^2 \tau}{D_{M,s}}\right)\right], \tau > 0, \xi = 0 \\ \text{stripping phase at the top of the boundary layer created due to reaction} \\ \theta_{M,s} = 1 + \dfrac{k_2}{k_1 - k_2}\exp\left(-\dfrac{k_1 \delta_s^2 \tau}{D_{M,s}}\right) + \dfrac{k_1}{k_2 - k_1}\exp\left(-\dfrac{k_2 \delta_s^2 \tau}{D_{M,s}}\right), \tau > 0, \xi = 1 \end{cases}$$

Within the membrane:

$$\begin{cases} \theta_{M,m} = 0, \tau = 0 \\ \text{membrane-feed phase interface, membrane side} \\ \theta_{M,m} = \dfrac{k_1}{k_2 - k_1}\left[\exp\left(-\dfrac{k_1 \delta_m^2 \tau}{D_{m,eff}}\right) - \exp\left(-\dfrac{k_2 \delta_m^2 \tau}{D_{m,eff}}\right)\right], \tau > 0, \xi = 0 \\ \text{membrane-stripping phase interface, membrane side} \\ \theta_{M,m} = 0, \tau > 0, \xi = 1 \end{cases}$$

where, $C_{M,f}$, $C_{M,m}$, and $C_{M,s}$ are the metal concentrations in the feed, membrane and stripping phases respectively, k_1 and k_2 are the first order rate constants from feed to membrane and membrane to stripping phase respectively. Figure 5.6 shows the concentration profile for the metal ion concentration.

Before deriving a solution for the above system, let us first understand how the concentration is derived at the boundary (92). The rate equation at the boundary layer for feed, membrane and stripping phase side can be written as,

$$\frac{dC_{M,f}}{dt} = -k_1 C_{M,f} \Rightarrow \frac{d\theta_{M,f}}{d\tau} = -\frac{k_1 \delta_f^2}{D_{M,f}} \theta_{M,f} \tag{5.22}$$

$$\therefore \theta_{M,f} = \exp\left(-\frac{k_1 \delta_f^2 \tau}{D_{M,f}}\right)$$

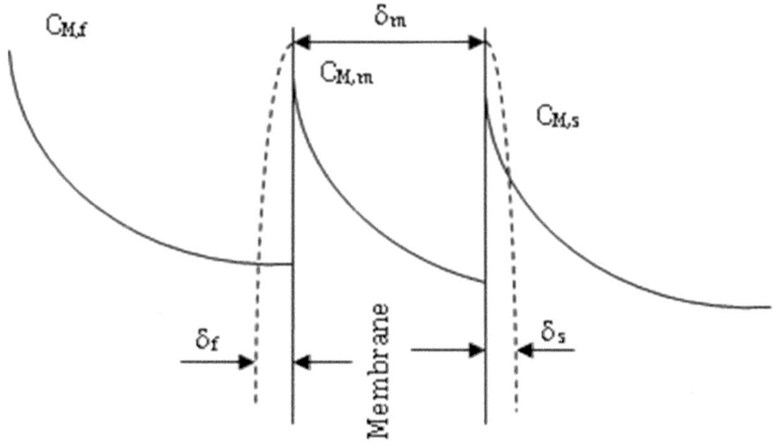

Fig. 5.6: Schematic of the mass transfer gradient across the membrane for SLM

$$\frac{dC_{M,m}}{dt} = k_1 C_{M,f} - k_2 C_{M,m} \Rightarrow \frac{d\theta_{M,m}}{d\tau} = \frac{k_1 \delta_m^2}{D_{M,m}} \theta_{M,f} - \frac{k_2 \delta_m^2}{D_{M,m}} \theta_{M,m}$$

$$\because \theta_{M,f} = \exp\left(-\frac{k_1 \delta_m^2 \tau}{D_{M,m}}\right) (\text{on membrane side})$$

$$\therefore \frac{d\theta_{M,m}}{d\tau} = \frac{k_1 \delta_m^2}{D_{M,m}} \exp\left(-\frac{k_1 \delta_m^2 \tau}{D_{M,m}}\right) - \frac{k_2 \delta_m^2}{D_{M,m}} \theta_{M,m}$$

$$\Rightarrow \frac{d\theta_{M,m}}{d\tau} + \frac{k_2 \delta_m^2}{D_{M,m}} \theta_{M,m} = \frac{k_1 \delta_m^2}{D_{M,m}} \exp\left(-\frac{k_1 \delta_m^2 \tau}{D_{M,m}}\right)$$

The integration factor is given by $\text{I.F.} = \int \frac{k_2 \delta_m^2}{D_{M,m}} d\tau = \frac{k_2 \delta_m^2 \tau}{D_{M,m}}$. After multiplying this integration factor on both sides of the differential equation and with this integration factor and integrating under the condition = $|$ 0, $\theta_{M,m} = 0$, we get,

$$\theta_{M,m} = \frac{k_1}{k_2 - k_1} \left[\exp\left(-\frac{k_1 \delta_m^2 \tau}{D_{m,eff}}\right) - \exp\left(-\frac{k_2 \delta_m^2 \tau}{D_{m,eff}}\right) \right]$$

Similarly, we also integrate for $\theta_{M,s}$. Now after applying the separation of variable techniques one has $\theta = X(\xi)\, T(\tau)$. Therefore unsteady state equations can be rewritten as, $XT' = TX''$. Hence,

$$\frac{T'}{T} = \frac{X''}{X} = -\lambda \tag{5.23}$$

Now we apply boundary conditions for all the cases with the feed, membrane and stripping phases at $\tau = 0$, $T = 0$. Hence, $T = \exp(-\lambda t) - 1$. Again with the second case, $X = C_1 \sin(\lambda \times \xi) + C_2 \cos(\lambda \times \xi)$.

Now based on the boundary conditions for the following phases the solution can be obtained as,

Feed side:

$$\theta_{M,f} = \exp\left(-\frac{k_1 \delta_f^2 \tau}{D_{M,f}}\right) \sum_n \left[\exp(-\lambda_n \tau) - 1\right] \frac{\sin(\lambda_n (1-\xi))}{\sin \lambda_n} \tag{5.24}$$

Membrane side:

$$\theta_{M,m} = \frac{k_1}{k_2 - k_1}\left[\exp\left(-\frac{k_1 \delta_m^2 \tau}{D_{m,eff}}\right) - \exp\left(-\frac{k_2 \delta_m^2 \tau}{D_{m,eff}}\right)\right] \sum_n \left[\exp(-\lambda_n \tau) - 1\right] \frac{\sin(\lambda_n (1-\xi))}{\sin \lambda_n} \tag{5.25}$$

Stripping side:

$$\theta_{M,s} = \sum_n \left[\exp(-\lambda_n \tau) - 1\right] \frac{\left[1 + \frac{k_2}{k_1 - k_2}\exp\left(-\frac{k_1 \delta_s^2 \tau}{D_{M,s}}\right) + \frac{k_1}{k_2 - k_1}\exp\left(-\frac{k_2 \delta_s^2 \tau}{D_{M,s}}\right)\right] + \frac{k_1}{k_2 - k_1}\left[\exp\left(-\frac{k_1 \delta_s^2 \tau}{D_{M,s}}\right) - \exp\left(-\frac{k_2 \delta_s^2 \tau}{D_{M,s}}\right)\right] \sin(\lambda_n (1-\xi))}{\sin \lambda_n} \tag{5.26}$$

5.4 Characterization of SLM Based on Organic Solvent and Membrane Support

One of the basic properties essential for the support of SLM is the hydrophobicity of the membrane, small pore size, appropriate tortuosity, and high porosity. Polyethersulfone (PES), polysulfone (PSF) and polyvinylidene fluoride (PVDF) are three basic types of membranes that are frequently used for the separation. In a study by Farah et al. (89), 0.45 micron PVDF and polytetrafluoroethylene (PTFE) membranes were used for the separation of drugs. They observed that the PTFE membrane has a more fibrous structure with higher porosity, which may indicate that it has more organic liquid entrapped inside the pores. However, with higher porosity, the loss of liquid is more pronounced and the stability is reduced compared to the more uniformly surfaced PVDF membrane support. In this case, polypropylene (PP) was found to be the most appropriate for the separation of pollutants as it is cheaper and compatible with a wide variety of organic solvents. Furthermore, it has a much smaller pore size with an increase in selectivity for the pollutants. The higher selectivity is attributed to the fact that there is a less chance of the matrix escaping into the acceptor (stripping) side (93). Additionally, the performance of SLM is not only limited by the support but also the nature of the organic solvent used. In a very recent study by Mubashir et al. (94) a detailed discussion was conducted on the organic solvent used in SLM preparation. These include volatile organic compounds (VOCs) (such as n-hexane, acetone), ionic liquids (ILs) (such as imidazolium, pyrrolidinium) and deep eutectic solvents (DES) (such as choline chloride and urea). VOC is the most conventional solvent used in SLM. However, the limitation with VOCs is their high volatility, which ultimately increases the cost of SLM. In contrast, ILs are much more stable with low vapor pressure. However, the major drawbacks of IL are its toxicity and cost. Recently, DES has emerged as a cheaper and almost non-toxic alternative with high chemical stability and low vapor pressure. DES is based on the combination of two compounds: a hydrogen bond acceptor and a hydrogen bond donor. Before delving deeper into DES, let's briefly define "eutectic". The eutectic point is the lowest temperature at which a solid melts directly into a liquid. According to literature (94), a nylon membrane with the DES dimethylamine nitrate ($DMANO_3$)/glycerol exhibits membrane stability for over 160 hours and optimal selectivity for separating acetylene from ethane. In another study (95), the DES N,N′-dimethyl urea (DMU)/mannose extracted nearly 47% of phosphorous after 24 hours of operation. More recently, DES based SLM has attracted considerable research interest, aiming to develop a more efficient and effective separation scheme.

5.5 Bulk Liquid Membrane (BLM)

In BLM, the feed and the receiving phase are separated by the organic liquid membrane through placing a water/organic immiscible barrier at the organic-aqueous interface (96). Figure 5.7 shows different types of configurations for BLM. The presence of a barrier at the interface depends on the design arrangement of the BLM. In (a) and (b) configurations there is no need for the barrier to be placed at the interface. On the contrary, in the other arrangements ((c) and (d)), there will be a membrane barrier that is actually placed at the interface (97). One of the typical disadvantages of BLM is the low transport rate due to the membrane's thickness, which restricts the passage of solutes through the liquid membrane. Although BLM offers membrane stability, it is not favourable for large scale operations because of the impeded solute transport. One of the primary considerations in selecting the solvent for the membrane phase in BLM is its viscosity. Higher viscosity results in lower diffusivity of solutes, primarily metals and reduces the mass transport rate across the membrane. Nonpolar organics are generally less viscous compared to polar solvents. This is due to the dipole moment within polar solvents, which manifests as high van der Waals forces of attraction. This characteristic is absent in apolar solvents (98). Table 5.2 details the viscosities of both polar and apolar organics used as membrane phases.

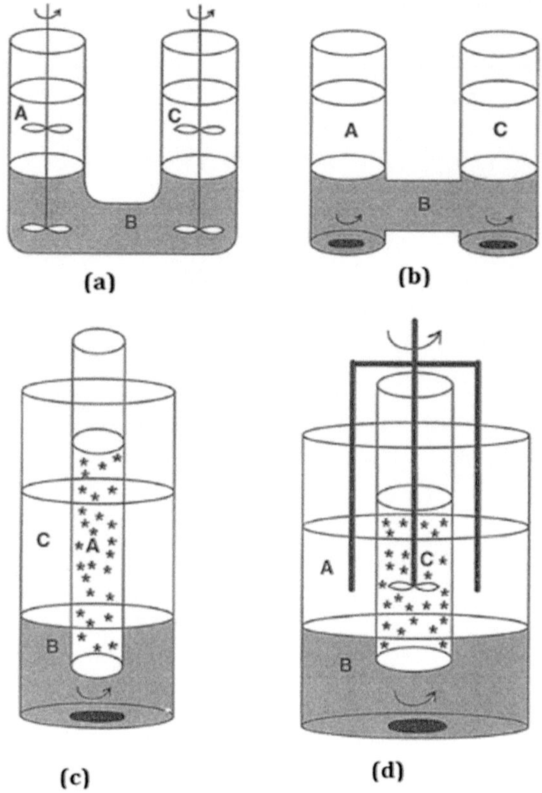

Fig. 5.7: (a)-(b): U-tube configuration, (c)-(d): tube within a shell configuration [Reproduced with permission from (99)]

Hence, one of the prime techniques which enhances mass transport is stirring the membrane phase. This increases the convective mass transfer coefficient, aiding mass transport across the solvent. The rate of mass transfer is higher here compared to mass transport by Fickian diffusion in nonstirring conditions. However, the efficiency of stirring becomes evident when a carrier within the organic phase accepts pollutants from the feed. This needs proper mixing within the organic phase and facilitates transport. Further, the acceptance of the pollutants from the feed and delivering them to the receiving phase is pH dependent. In one example Memon et al. (100) wanted to transport methyl red dye from the feed phase (pH = 2) to the receiving phase (pH = 12). They used the facilitated transport mechanism, where a carrier calyx[6]arene based derivative is mixed with the organic chloroform and the carrier has an affinity for methyl red. The carrier has aromatic rings connected to each other through π-π interactions, creating a cavity accommodating the dye through complexation. In this case, complex formation occurs through proton arrest within the cavity, and the presence of

Table 5.2: Viscosities of different solvents used for the fabrication of BLM [Reproduced with permission from (98)]

Solvent type	**Viscosities (mPa.s)**
Nonpolar solvent	
Hexane	0.30 (25°C)
Heptane	0.376 (27 °C)
Octane	0.51 (25 °C)
Decane	0.86 (25 °C)
Undecane	1.09 (100 °C)
Dodecane	1.37 (25 °C)
Cyclohexane	0.887 (25 °C)
Kerosene	1.64 (27 °C)
Dichloromethane	0.406 (25 °C)
1,2-Dichloroethane	0.776 (25 °C)
Chloroform	0.53 (27 °C)
Carbon tetrachloride	0.97 (20 °C)
Benzene	0.601 (27 °C)
Toluene	0.55 (27 °C)
Xylene	0.605 (25 °C)
Nonpolar	
1-Heptanol	5.937 (25 °C)
1-Octanol	7.598 (25 °C)
1-Decanol	11.797 (25 °C)
Cyclohexanone	2.02 (25 °C)
Tributylphosphate	3.8 (20 °C)
Tri-n-Octylamine	8.325 (25 °C)
Trioctylphosphine oxide	15 (55 °C)
Di-2-ethylhexylphosphoric acid	21.22 (30 °C)
Quinoline	3.36 (25 °C)
Water	0.89 (27 °C)

protons at acidic pH makes the methyl red bind. Conversely, at alkaline pH, the protons are released from the cavity, decomplexing the dye. Therefore, pH is a crucial parameter that affects the separation extent of pollutants using BLM and needs careful maintenance for effective removal. The effect of temperature on the efficiency of BLM can be observed. As temperature increases, the separation efficiency of the bulk liquid membrane also increases. This is due to the reduction in membrane viscosity with higher temperatures, as explained by the Stokes-

Einstein equation (equation 5.27) (101). As a result, transportation through the organic liquid membrane becomes spontaneous.

$$D_M = \frac{k_B T}{\pi r_s \mu c} \tag{5.27}$$

The removal of pollutants with BLM is carrier facilitated and depends entirely on the concentration gradient across the membrane. Therefore, the feed-to-receiving phase volume ratio is a crucial factor in BLM design. If this ratio changes, the separation efficiency will be affected. As the ratio increases, the receiving phase will approach saturation with pollutants, reducing the concentration gradient and decreasing the diffusive mass transport from the membrane to the receiving phase. A reduction in the concentration gradient across the membrane-receiving phase interface will decrease the mass transfer from the membrane to the receiving phase. Conversely, the opposite scenario is observed with a low ratio. However, if all the operating parameters such as stirring speed, pH, temperature and the ratio are considered together, they will be found to be interrelated. Stirring ensures proper carrier mixing, facilitating transport. Temperature promotes transfer by reducing viscosity. pH ensures decomplexation and effective mass transfer from feed to membrane to receiving phase. The ratio highlights the propensity of diffusion and, consequently, the transfer of pollutants. To achieve optimal operating conditions, it is essential to properly design BLM. The following subsection provides an elaboration on the mathematics of the transport mechanism for BLM to implement an appropriate module design.

5.5.1 Transport Mechanism in BLM

As discussed previously, the effectiveness of BLM is assessed based on several parameters. Including all parameters in a realistic mathematical model for BLM would be difficult to solve due to high nonlinearity and mutual inclusiveness. Therefore, the following assumptions have been considered to simplify the mathematical model.

- There will be pollutant transport at the feed-membrane and membrane-receiving phase interface.
- There will be proper mixing at constant temperature.
- The physical and transport properties are constant.
- The membrane is completely immiscible with the feed and receiving phase.
- There will be no short circuit between the feed and the receiving phase.
- The partition coefficient includes the pollutants and the pollutant-carrier complex.
- The concentration of pollutants in the membrane and receiving phase are zero at the very beginning of the operations.

Now let's consider the carrier facilitated transport given in equation 5.15.

$$\left.\begin{aligned} &M^{2+}(\text{aq, feed}) + 2HR(\text{org}) \underset{k_{-1}}{\overset{k_1}{\rightleftharpoons}} R_2M(\text{org}) + 2H^+(\text{aq, feed}) \\ &2H^+(\text{aq, stripping}) + R_2M(\text{org}) \underset{k_{-2}}{\overset{k_2}{\rightleftharpoons}} M^{2+}(\text{aq, stripping}) + 2HR(\text{org}) \end{aligned}\right\} \quad (5.28)$$

The concentration of metals, previously referred to as pollutants, is now expressed as a fraction of the initial feed concentration (equation 5.29).

$$\left.\begin{aligned} R_{M,f} &= \frac{C_{M,f}}{C_{M,f0}} \\ R_{M,m} &= \frac{C_{M,m}}{C_{M,f0}} \\ R_{M,s} &= \frac{C_{M,s}}{C_{M,f0}} \end{aligned}\right\} \quad (5.29)$$

Thus, the fluxes for the metal from feed side to membrane phase and from the membrane phase to the stripping (receiving) phase are given by equations 5.30, 5.31 and 5.32, respectively (102). The reactive terms associated with the binding of pollutants to the carrier are incorporated into the following mass balance equations.

$$\left.\begin{aligned} \frac{dR_{M,f}}{dt} &= k_{-1}R_{M,m} - k_1 R_{M,f} \\ J_{M,f} &= \frac{1}{a}\left(k_{-1}R_{M,m} - k_1 R_{M,f}\right) \end{aligned}\right\} \quad (5.30)$$

$$\left.\begin{aligned} \frac{dR_{M,m}}{dt} &= k_1 R_{M,f} + k_{-2}R_{M,s} - \left(k_{-1} + k_2\right)R_{M,m} \\ J_{M,m} &= \frac{1}{a}\left\{k_1 R_{M,f} + k_{-2}R_{M,s} - \left(k_{-1} + k_2\right)R_{M,m}\right\} \end{aligned}\right\} \quad (5.31)$$

$$\left.\begin{aligned} \frac{dR_{M,s}}{dt} &= k_2 R_{M,m} - k_{-2}R_{M,s} \\ J_{M,s} &= \frac{1}{a}\left(k_2 R_{M,m} - k_{-2}R_{M,s}\right) \end{aligned}\right\} \quad (5.32)$$

Considering the reaction given in equation 5.28, it is desirable for the feed pH to be higher and the stripping phase to be lower. A lower pH in the stripping phase causes protonation, preventing the backward reaction, thus k_{-2} can be neglected. Similarly, with a higher pH on the feed side, the backward reaction rate k_{-1} can be neglected. Therefore, equations 5.30 to 5.32 can be modified and represented as 5.33 through 5.35.

$$\left.\begin{aligned}\frac{dR_{M,f}}{dt} &= -k_1 R_{M,f}\\ J_{M,f} &= \frac{1}{a}\left(-k_1 R_{M,f}\right)\end{aligned}\right\} \tag{5.33}$$

$$\left.\begin{aligned}\frac{dR_{M,m}}{dt} &= k_1 R_{M,f} - k_2 R_{M,m}\\ J_{M,m} &= \frac{1}{a}\left(k_1 R_{M,f} - k_2 R_{M,m}\right)\end{aligned}\right\} \tag{5.34}$$

$$\left.\begin{aligned}\frac{dR_{M,s}}{dt} &= k_2 R_{M,m}\\ J_{M,s} &= \frac{1}{a}\left(k_2 R_{M,m}\right)\end{aligned}\right\} \tag{5.35}$$

The boundary conditions for the above equations can be set as, at $t = 0$, $R_{M,f} = 1$, $R_{M,m} = 0$ and $R_{M,s} = 0$. Applying these initial conditions, the equations can be solved to evaluate the values for the rejection factors.

$$\left.\begin{aligned}R_{M,f} &= \exp\left(-k_1 t\right)\\ J_{M,f} &= \frac{1}{a}\left(-k_1 \exp\left(-k_1 t\right)\right)\end{aligned}\right\} \tag{5.36}$$

Using the value of $R_{M,f}$ obtained from equation 5.36 in equation 5.34, $R_{M,m}$ is evaluated as given in equation 5.37

$$\left.\begin{aligned}R_{M,m} &= k_1\left[\frac{\exp\left(k_1 t\right)}{k_1\left(k_1+k_2\right)} + \frac{\exp\left(-k_2 t\right)}{k_2\left(k_1+k_2\right)} - \frac{1}{k_1 k_2}\right]\\ J_{M,m} &= \frac{1}{a}\left(k_1\exp\left(-k_1 t\right) - k_1 k_2\left[\frac{\exp\left(k_1 t\right)}{k_1\left(k_1+k_2\right)} + \frac{\exp\left(-k_2 t\right)}{k_2\left(k_1+k_2\right)} - \frac{1}{k_1 k_2}\right]\right)\end{aligned}\right\} \tag{5.37}$$

$$J_{M,s} = \frac{1}{a}\left(k_1 k_2\left[\frac{\exp\left(k_1 t\right)}{k_1\left(k_1+k_2\right)} + \frac{\exp\left(-k_2 t\right)}{k_2\left(k_1+k_2\right)} - \frac{1}{k_1 k_2}\right]\right) \tag{5.38}$$

5.5.2 BLM Applications in Analytical Process

One of the typical applications with BLM is the preconcentration of analytes, which is quite economical compared to other techniques. Let's start with some literatures surveyed by several researchers to prepare samples for analysis. Candela et al. (103) preconcentrated uranium (VI) using BLM, with kerosene as the membrane phase containing bis(2-ethylhexyl) phosphoric acid (D2EHPA) as the carrier dissolved in the organic phase. Uranium is a radioactive substance that arises from various anthropogenic activities, such as mining, mineral processing, coal combustion. It is evident that uranium drainage poses both radiological and chemical toxicity. This necessitates the determination of uranium through analysis using molecular spectrophotometry, ICP mass spectrophotometry, voltammetry, etc. According to a report, uranium concentration in shallow ground water ranges from 0 to 532 ppb across India (104). In the USA, particularly in the Grand Canyon region, it ranges from 1 to 100 ppb due to uranium mines found in geological features called "Breccia pipes" (105, 106). Detecting trace levels of uranium (<100 ppb) is a common task. The transportation of uranium is described by the reaction in equation 5.39. Here, the carrier D2EHPA, an acidic agent with a highly hydrophobic anion, forms a metal complex by adopting metal cations and transporting them from the aqueous to the organic phase.

$$UO_2^{2+}(\text{aq, feed}) + 2H_2R_2(\text{org})(\text{D2EHPA}) \underset{k_{-1}}{\overset{k_1}{\rightleftharpoons}} UO_2R_2(HR)_2(\text{org}) + 2H^+(\text{aq, feed}) \tag{5.39}$$

The preconcentration factor (PF) is given by the equation 5.40.

$$PF = \frac{UO_2^{2+}(\text{aq, stripping})}{UO_2^{2+}(\text{aq, feed})} \tag{5.40}$$

For the same reason described earlier, the pH of the stripping phase was lowered using a mixture of citric acid and phosphoric acid in varying proportions, along with different quantities of D2EHPA. It was observed that the PF was higher with a substantial amount of D2EHPA, as described for system 1 in Table 5.3.

Table 5.3: Chemical composition of solvent extraction system for uranium in BLM [Reproduced with permission from (103)]

	D2EHPA (mol/L)	Citric acid (mol/L)	Phosphoric acid (mol/L)
System 1	0.004	0.450	0.646
System 2	0.001	0.150	0.407

Pirmoradi and Ashrafizadeh (107) separated nitrate from water using BLM. Nitrate's toxicity, particularly for infants, can cause "blue baby syndrome". Due to its high solubility and stability, conventional technologies like filtration struggle to remove nitrate. Effective technologies include ion exchange, electrodialysis, reverse osmosis and biological treatment. However, ion exchange requires resin bed regeneration, electrodialysis and reverse osmosis are costly, and biological treatment demands rigorous downstream operations to remove carbon resources, nutrients and pathogens. Liquid membrane technology offers an alternative approach due to its high selectivity and low energy consumption. dimethyldioctadecylammonium chloride (DODMAC) was used as a carrier reagent for nitrate removal. The carrier was dissolved in a solution of 10% chloroform and 90% normal hexane, serving as the organic membrane phase. The membrane phase, due to its low density, is placed at the top of the aqueous phase (inverted shell configuration as shown in Figure 5.7). Maximum PF was observed with a carrier concentration of 0.0075 M. Instead of using PF to represent the efficacy of the BLM, it's better to evaluate it by the percentage of recovery (R%, equation 5.41), which is around 99%.

$$R\% = \left(1 - \frac{UO_2^{2+}\left(\text{aq, stripping}\right)}{UO_2^{2+}\left(\text{aq, feed}\right)}\right) \times 100 \qquad (5.41)$$

Table 5.4: shows the metal recovery percentage for different solvents using BLM

Solvent system (carrier)	Metal ions	Recovery percentage	Reference
Nitrobenzene (Kryptofix5)	Fe^{3+}	98.6%	(108)
	Zn^{2+}	98.5%	
	Ag^{1+}	87.7%	
Chloroform (Kryptofix5)	Fe^{3+}	99.5%	
	Zn^{2+}	98.2%	
	Ag^{1+}	35.6%	
Dichloromethane (Kryptofix5)	Zn^{2+}	97.1%	
	Ag^{1+}	40.4%	
1,2-Dichloroethane (Kryptofix5)	Ag^{1+}	82.8%	
Nitrobenzene (DCH18C6)	Ag^{1+}	87.0%	(109)
	Pb^{2+}	54.5%	
Chloroform (DCH18C6)	Ag^{1+}	85.3%	
	Pb^{2+}	48.9%	
Dichloromethane (DCH18C6)	Ag^{1+}	66.5%	
	Pb^{2+}	60.0%	
1,2-Dichloroethane (DCH18C6)	Ag^{1+}	46.3%	

Kryptofix5: 1,13-bis(8-quinolyl)-1,4,7,10,13-pentaoxatridecane
DCH18C6: Dicylohexyl-18-crown-6

5.6 Summary

- A brief elaboration on liquid membrane technology.
- Elaboration on ELM, SLM and BLM.
- Solvent system design for liquid membrane.
- Carrier selection for facilitated transport with SLM and BLM.
- Transport mechanism and its modeling for ELM, SLM and BLM.
- Characterization of SLM.
- Applications of different types of liquid membranes.

CHAPTER

6

Dialysis and Extracorporeal Membrane Oxygenator (ECMO)

6.1 Introduction

The word 'dialysis' was first coined in the year 1861 by a professor of chemistry named Thomas Graham at Anderson's University. He first observed dialysis as the diffusion phenomenon of the crystalloids through albumin coated vegetable parchment, acting as a semi-permeable membrane. He then used this concept for extracting urea from the urine samples. In 1913, scientists Abel, Rowntree, and Turner from John Hopkins Medical College developed the first artificial kidney, using treated parchment to synthesize the membrane. Before the experiment, sodium salicylate was introduced as a marker to mimic renal failure, where toxins and waste accumulate in the body. This marker was subsequently removed through the dialyzer which consisted of 16 collodion tubes, each about 40 cm length and 8 mm in diameter, much larger than current commercial models. Due to the delicacy of collodion, other natural materials were tested as separation membranes. In 1943, WJ Kolff and H Berk fabricated the first practical haemodialysis machine using a 30–40 m cellophane tubing fixed inside a 100-liter stationary tank. Artificial blood containing urea in a concentration 400 mg/100 ml was treated by the membrane in 0.5 hours, eliminating the urea (110). Dr. WJ Kolff, known as the "Father of Dialysis", emphasized the mechanical strength of membrane materials over separation performance and permeability during that time.

The first clinical trial of kidney dialysis was conducted by Dr. Nils Alwall at the Department of Internal Medicine in Lund University Hospital. The dialysis machine used in this trial consisted of a glass cylindrical container with a stainless steel wire mesh and a flattened cellophane tube. Contaminants were separated from blood samples by passing the blood through this casing. Later, on March 9, 1960, Dr. Belding Scribner, Dr. Wayne Quinton, and Dr. David Dillard performed the first clinical trial on a chronic dialysis patient by using a Teflon R tubing shunt. This teflon tubing was non-reactive with tissue and significantly safer. The dialysis process lasted for 11 hours. In 1961, Dr. Scribner and Norwegian urologist Fred Kiil had developed a less resistant flat plate dialyser fitted with Cuprophane, a cellulose based membrane (111). In 1962, Cimino and Brescia introduced the "arteriovenous fistula" concept, creating a veno-venous access for dialysis. A blood pump and a sphygmomanometer were used to dilate an accessible forearm vein, and purified blood was returned via another vein in the ankle (112, 113). Numerous advancements have been made in the fabrication of dialysis machines. Recently, research has focused on designing artificial transplantable kidneys to alleviate the pain and strain of dialysis. Dr. Shuvo Roy, a bioengineer and professor at UCSF, is developing an artificial implantable kidney compatible with human kidney cells, without disturbing the immune system.

6.2 Overview on Dialysis Process

Before delving into an in-depth analysis of dialysis let's understand its fundamentals. Dialysis is a separation process used to remove waste toxic products and excess water from the blood, serving as an artificial substitute for kidney function in cases of renal failure. Though it doesn't fully replicate kidney functions, it significantly aids through diffusion and ultrafiltration. Dialysis is performed in chronic renal failure (CRF) cases when the glomerular filtration rate drops below 15 ml/min/1.73 m^2. Chronic renal failure (CRF) is a condition where kidney function is lost over months or years. Diagnosis involves measuring serum creatinine levels to determine glomerular filtration rate (GFR), with five stages of CRF based on GFR. Dialysis is commonly performed in stage 5 (end stage renal disease (ESRD)) when GFR is less than 15 ml/min/1.73 m^2, to remove accumulated toxins. The kidneys' primary functions include fluid, acid/base, and electrolyte balance all of which dialysis replicates using a countercurrent flow between blood and dialysate through a hollow-fibre channel. When kidney function drops below survival levels, external systems are needed to support renal functions. Common treatments include hemodialysis, peritoneal dialysis, hemoadsorption, hemodiafiltration, and hemofiltration, with hemodialysis being the most accepted. The artificial kidney device which is utilized in hemodialysis process, termed "hemodialyzer" or "dialyzer", is equipped with a membrane to remove toxins, waste products and excess water from the bloodstream. Additionally, it helps balance electrolytes and modifies acidic pH.

A key complexity of dialysis is that its functions depend on the dialysis membrane material. Membrane flux significantly affects the permeation of solutes with varying molecular weights. Additionally, it's crucial to ensure no harmful reaction occur between the blood immune cells and the membrane material when in contact. According to Mumtaz et al., 2010 (114), synthetic membranes with high porosity showed better passage of β_2-microglobulin (a kidney-excreted component with a molecular weight of 11,900 Dalton) compared to cellulose based membranes. However, low-porosity membranes increased β_2-microglobulin concentration, potentially raising the mortality rate among renal failure patients. The removal of other uremic toxins, such as protein carbonyls (PCO) and advanced oxidation protein products (AOPP), further complicates the process. Previous research indicates that AOPP removal remains unaffected even with high flux membranes. This suggests that using low flux membranes, which allow toxin accumulation during chronic haemodialysis, could reduce mortality rates. In the following sections, we'll delve into the dialysis membranes, providing insight into the synergistic interaction of physics and chemistry with blood and the membrane.

6.2.1 Basic Principle of Dialysis

Diffusion is the movement of solute and solvent molecules due to concentration differences. Diffusion dialysis,a membrane process, separates various solutes through this diffusion across the membrane. It is typically used when changes in solute composition or concentration are needed. In biochemical operations, dialysis is used to adjust the concentration of salts and small molecules in protein solutions. The main goal is to reduce the concentration of these solutes, though the composition or concentration of the solution can also be modified in other ways. Diffusion dialysis relies on diffusion, where solute mobility between two liquid spaces is mainly restricted by size. In rare cases, charge or polarity may also restrict diffusion. Size restriction is achieved using a porous semipermeable dialysis membrane. The dialysis membrane permits only particles below a certain size. In biochemical labs, it's often a transparent tube tied at both ends, known as a dialysis bag. Commonly used membranes for diffusion dialysis include collodion, cellophane,and parchment paper. While suitable for industrial applications like alkali recovery in the rayon industry, these membranes were less used industrially due to low mechanical stability. This limitation was resolved with the development of synthetic polymeric membranes.

Let's understand the phenomenon with an example. A solution of volume V1 is placed in a dialysis bag. The bag is then immersed in a dialysis solution of volume V2, which is continuously stirred at a slow rate. Stirring aids the diffusion of solutes through the bag membrane, helping to reach equilibrium between solute concentrations in the two liquid spaces. If the volume difference between the two liquid spaces is large ($V2 >> V1$, such as $V2 = 200$ L and $V1 = 2$ L, a 100 fold difference), the initial equilibrium will cause significant dilution of the small solutes present originally in the bag. The concentration change of small solutes occurs by a factor of $V1/(V1 + V2)$, which is quite small ($<<1$), while the concentration change in the outer solution is nearly 1 ($(V2)/(V1 + V2)$). However, the concentration of solutes that cannot penetrate the membrane remains almost unchanged.

6.2.2 Hemodialysis and Its Mechanism

The principle of haemodialysis is similar to other dialysis methods, involving the diffusion of solutes through a semipermeable membrane. Hemodialysis removes toxic wastes and excess water from blood using a hemodialyzer device, which is equipped with a semipermeable membrane. The separation process creates a counter-current flow gradient, with blood flowing in one direction and the hemodialyzer fluid in the opposite direction. The primary mechanism in dialysis is diffusion, where solute particles diffuse through the semipermeable membrane. Metabolic waste and toxic products like creatinine and urea are pulled down the concentration gradient from the blood into the dialysate. Dialysate, also known as dialysis fluid is a solution of deionized water and electrolytes (sodium bicarbonate ($NaHCO_3$), sodium chloride (NaCl)). During diffusion, the rate at which toxins diffuse into the dialysate largely depends on the size of the solute particles. The smaller the size of the particles, the faster they diffuse through the semipermeable membrane and vice-versa. An arteriovenous shunt is created by connecting arteries carrying oxygenated blood from the heart to a vein, making the vein strong enough to withstand multiple punctures. Regular monitoring of pressure is essential during the dialysis process. The schematic representation of the dialyzer is provided in Figure 6.1. Dialyser membranes come in various pore sizes: membranes with small pores are called 'low-flux' and those with larger pores are termed 'high-flux' dialysers. For example, larger molecules like β2-microglobulin are not removed by low-flux dialysers. The current trend is to use high-flux dialysers. However, high-flux dialysers require advanced dialysis machines and high quality dialysis solutions to control removal rates and prevent backflow of impurities into the patient's body across the membrane.

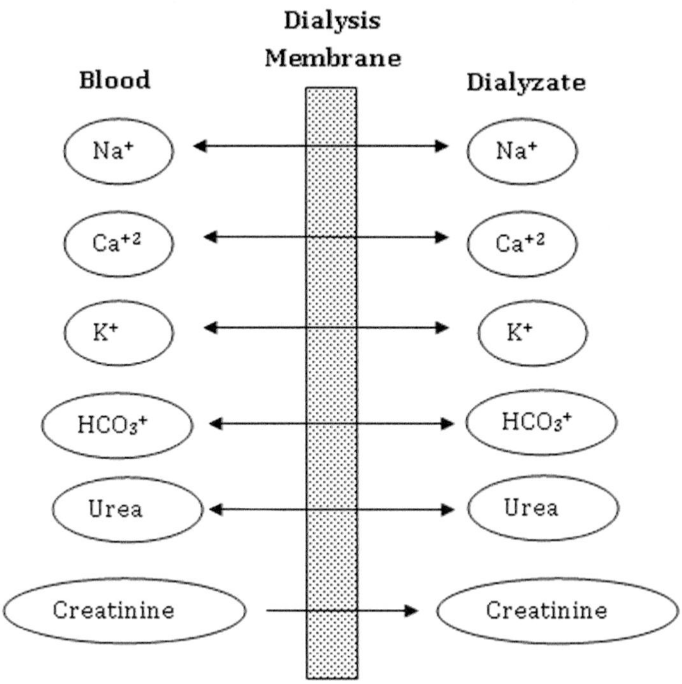

Fig. 6.1: Schematic representation of the transport of solutes from blood to dialyzate in countercurrent dialysis operation

6.2.2.1 Law of Diffusion

In the dialysis process, two fluids typically flow on opposite sides of the membrane barrier in a counter-current direction. As a result solutes present in higher concentration move across the membrane due to the concentration gradient following, Fick's 1st law of diffusion.

$$J_{Ax} = -D_{Ax}\frac{\delta C_A}{\delta x} \tag{6.1}$$

where, x denotes the direction of diffusion (m), C_A is the concentration of solute A (kg/m^3), J_{Ax} is the mass flux of solute A in the x direction (kg/m^2.s), and D_{Ax} is the diffusion coefficient of solute A in the x direction (m^2/s).

Therefore, dialysis is also utilized in various industrial and laboratory situations as an effective separation technique. Assuming C_{A0} and C_{AL} are the concentrations of solute A at distances $x = 0$ and $x = L$, respectively (as shown in Figure 6.2), integrating Eq. 6.1, yields the following equation:

$$J_{Ax} = -\frac{D_{Ax}}{L}(C_{A0} - C_{AL}) \equiv k_M (C_{A0} - C_{AL}) \tag{6.2}$$

where, k_M represents the membrane permeability (m/s). From Eq. 6.2, it is evident that the diffusion rate varies directly with the concentration gradient (ΔC_A) between the two sides of the membrane barrier.

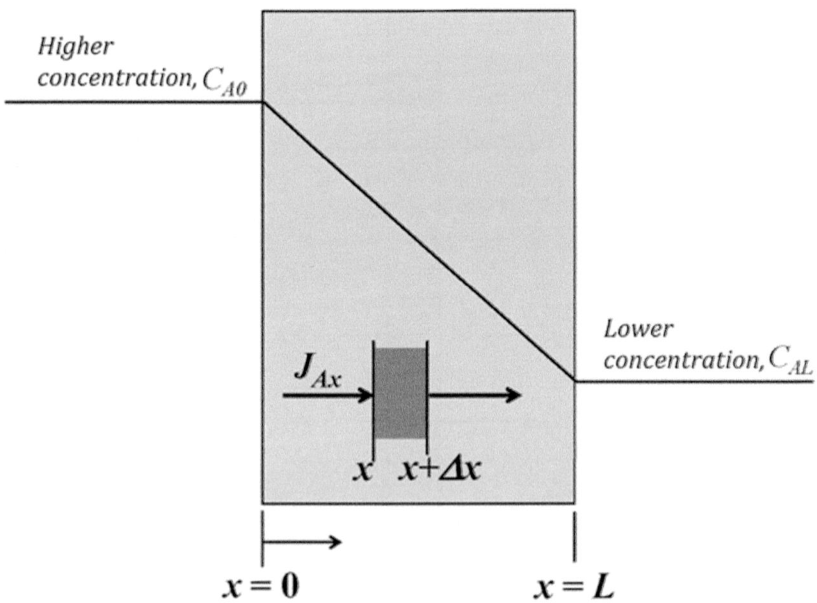

Fig. 6.2: Variation of concentration of solute in the dialysis process

6.3 Dialysis Membrane

The primary objective of using a dialysis membrane is to purify blood by removing accumulated uremic solutes due to kidney failure and restoring reduced components like bicarbonates. The permeability of the membrane fitted in a hemodialyzer is crucial for effective renal treatment. Various materials, both natural and synthetic polymers, have been investigated for membrane use. The following sections will discuss important dialysis membranes. The key issue in selecting a dialysis membrane is its biocompatibility, ensuring no inflammatory problems and high water permeability. The membrane material significantly impacts cytokine production, potentially causing inflammation. The main complexity in haemodialysis process is the limited sterility of the dialysate, which can lead to blood contamination.

Cuprophan® is a widely used dialysis membrane made cuprammonium rayon, which is created by dissolving cellulose in cuprammonium solution. It was first introduced commercially by Enka Co. in West Germany, then by Membrana, a Polypore Co. in Germany. Another cuprammonium rayon-based membrane, Bemberg®, with a similar structure to Cuprophan®, was produced by Asahi-Kasei Co. in Tokyo and later by Terumo. These cellulose-cast membranes, also known as regenerated cellulosic (RC) membranes, have limitations. Cuprophan® membrane, a cellulose based membrane, shows high affinity for the lipid A6 structure present in lipopolysaccharides (LPS), limiting its biocompatibility. To enhance biocompatibility, chemical modifications can be made by replacing hydroxyl groups with acetate groups. These membranes are classified as cellulose acetate, cellulose diacetate, and cellulose triacetate based on the number of acetate group replacements within the cellulose backbone. Although the commercial availability of RC membranes has decreased in recent decades, cellulose acetate, cellulose diacetate, and cellulose triacetate membranes still hold a significant market share due to their high solute permeability, hydraulic permeability, and biocompatibility.

The first polymeric dialysis membrane, AN-69® was developed by Rhône-Poulenc in France in 1969. The flat sheet AN-69® membrane dialyzer was commercially known as RP-6®. RP-6® was the first dialyzer that could be sterilized using gamma-ray irradiation. Although the production of AN-69® membranes has changed hands from Rhône-Poulenc to Hospal, Baxter, and Gambro, AN-69® membrane dialyzers are still used worldwide. These membranes are promising for acute kidney injury (AKI) therapy due to their strong adsorption of substances, including inflammatory cytokines.

In 1967, chemical engineers Stuart and Lipps from MIT developed the first cellulosic hollow-fiber membrane dialyzer. However, the product was first commercially introduced in 1972 by Cordis-Dow Co. (Miami, U.S.A.). These hollow-fiber dialyzers are structurally similar to multi-tube heat exchangers offer advantages such as compactness and a larger surface area, leading to their wide acceptance. Conversely, the polymethylmethacrylate (PMMA) membrane dialyzer was first introduced by Toray Co. in Tokyo, Japan, and can be sterilized by gamma-ray irradiation. Since the early 1980's, several synthetic polymeric membranes have been introduced to the market, improving solute permeability, hydraulic permeability, and biocompatibility. Even today, these polymeric membranes are the leading choice for dialysis treatment. Among them, polyethersulfone (PES), polysulfone (PSf) and polyarylethersulfone (PAES) membranes hold the largest market share.

Polysulfone (polymeric), Cuprophan® (cellulosic), and AN69® (copolymer of AN and sodium methallyl sulfonate) membranes are the most commonly utilized membranes for dialysis. These membranes share a key attribute: hydrophilicity, which gives them a hydrogel structure and high water permeability. Hydrogels are three-dimensional networks capable of absorbing and holding water. Due to the higher affinity of water molecules for both hydrophilic and hydrophobic tails, the interfacial free energy is reduced. Consequently, the surface shows a low tendency for adhering cells and proteins during the dialysis process. Here, a concise comparison of these three commonly used dialysis membranes is presented with reference to toxin elimination. The AN69 membrane is composed of acrylonitrile and sulfonate groups (Figure 6.3). The sulfonate group attracts water molecules, developing a 3-D hydrogel structure on the membrane. This improves the membrane's biocompatibility and permeation flux. Specifically, the AN69 membrane maximizes β_2-microglobulin elimination due to the adsorption of the protein molecules on the sulfonated surface (115). However, the adsorption of LPS is lower in the cuprophane membrane compared to the polysulfone and AN69 membranes (116). The elution of LPS from the cuprophane membrane washing solvent was only 2%, slightly lower than the 3.5% elution from the AN69 membrane. In contrast, the polysulfone membrane washing solvent showed a much higher elution of 9%. Thus, it can be affirmed that the AN69 membrane is preferable than the other two membranes due to its high permeation flux and lower infection risk from reduced LPS adhesion. Figure 6.4 depicts the schematic for the dialysis process.

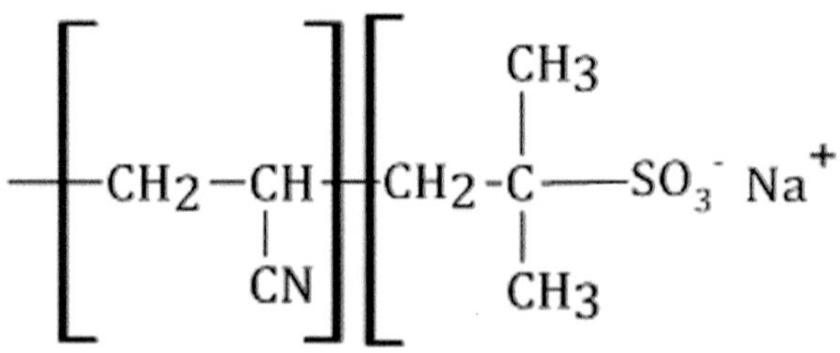

Fig. 6.3: Groups on the surface of AN69 dialysis membrane which is a copolymer of acrylonitrile (AN) and sodium methallyl sulfonate

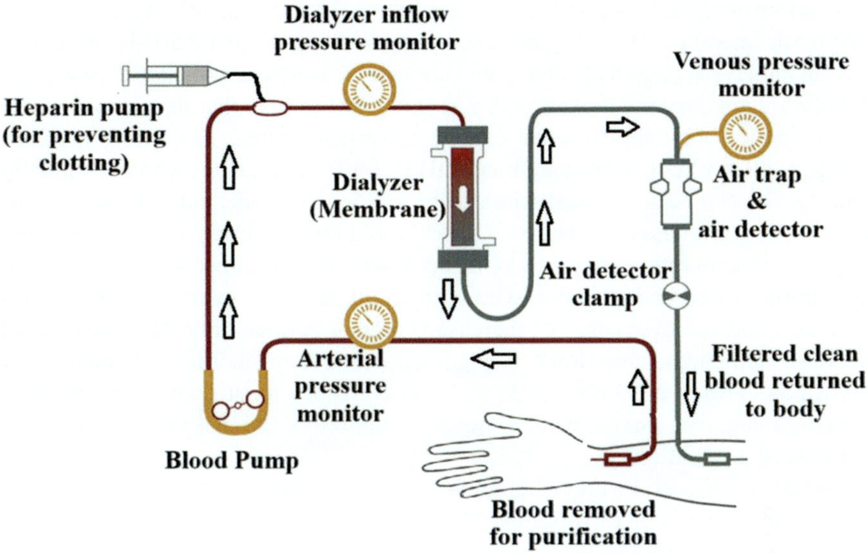

Fig. 6.4: Schematic of the dialysis process (117)

6.4 Biological Consideration of the Dialysis Membrane

Biocompatibility of dialysis membranes is a prime biological consideration. Dialyzers are used for four hours per session, thrice a week. Due to repeated applications, there is a high risk of undesired side effects, including chronic inflammation.

Until the 1970s, RC dialysis membranes dominated the global market, and and they were eventually replaced by synthetic polymeric dialysis membranes. Transient leukopenia, a common bioincompatible condition, was observed, characterized by a sudden decline in leukocytes 15-30 minutes after the start of dialysis treatment. In the 1970s, dialyzer's reprocessing included four steps: rinsing, cleaning, performance testing, and sterilization. Bioincompatibility was often observed during the first use of a dialyzer, commonly referred to as the "first use syndrome". According to Craddock et al. (1977), the use of RC dialysis membranes leads to complement activation, causing transient accumulation of leukocytes in the lungs and blood vessels (118). The backbone of RC contains three hydroxyl groups, which are linked to undesirable complement activation. To address this, one, two, or three acetate groups are introduced to replace the hydroxyl groups, resulting in the synthesis of CA, CDA, and CTA, respectively. These semi-synthesized cellulosic-based membranes not only improved biocompatibility but also enhanced water and solute permeability.

In this paragraph, we will discuss enhancing the biocompatibility of synthetic polymeric membranes. The glomerular basement membrane (GBM) present in human kidneys is negatively charged. Although the synthetic polymeric membrane AN-69® is also negatively charged, it is important to note that its use may cause anaphylactic shock shortly after dialysis treatment begins. The strong negative charge (–70 mV) converts Hageman (XII) factor to XIIa factor, which eventually induces the synthesis of bradykinin from kininogen. Under normal conditions, kininase II deactivates bradykinin. However, if a renal failure patient is taking an angiotensin-converting enzyme (ACE) inhibitor, kininase II becomes deactivated. This initiates a cascade reaction with bradykinin, leading to various health issues such as blood vessel expansion, increased vascular permeability, decreased blood pressure, and undesirable shock during treatment. This phenomenon is commonly known as "negative charge syndrome" (NCS). Since most dialysis membranes possess a negative charge, prescribing ACE inhibitors to renal failure patients is considered contradictory.

Furthermore, efforts are being to increase membrane biocompatibility through surface improvement techniques. Hemophan® is one such surface-modified membrane, is synthesized by incorporating a positively charged component, diethylaminoethyl (DEAE), into RC membrane (Membrana, Germany). Introducing a small amount of DEAE significantly restricted complement activation compared to the entire cellulose content. However, the limitation of the Hemophan® membrane is that it adsorbs heparin, triggering blood coagulation problems. Consequently, its commercial production was stopped. Additionally, RC membranes were coated with vitamin E to make them anti-oxidative (Terumo, Tokyo, Japan). This technique was later applied to PSf membranes and is still commercially manufactured by Asahi Kasei Medical Co., Tokyo, Japan.

Polysulfone (PSf) and similar membranes dominate the global dialysis membrane market. Due to their hydrophobicity, polyvinylpyrrolidone (PVP) is commonly used as a hydrophilic agent. Recently, PVP, used as a plasma supplement in medicines, has been linked to cases of anaphylactic shock, possibly due to its incorporation in the membrane. A clinical investigation used dialyzers with PSf and PEPA membranes containing varying amounts of PVP (119). The concentration profile of C3a over time, shown in Figure 6.5, indirectly measured complement activation. The PSf membrane with high PVP(+++) exhibited three times the C3a concentration 15 minutes after dialysis treatment began. The increase of C3a was slightly less during the first application of PEPA with PVP(+). Peak concentrations were nearly half or lower during weeks two to five. After five weeks of using PEPA with PVP(+), the peak concentration tripled with subsequent PSf use. Another clinical investigation showed that the highest C3a elevation occurred with PSf and PVP(+++), followed by PEPA with PVP(+), PVP(++), and PVP(–). C3a concentration elevation depended on the amount of PVP hydrophilic agent in the membrane rather than the membrane material itself. Therefore, PVP is not a suitable hydrophilic agent considering biocompatibility issues.

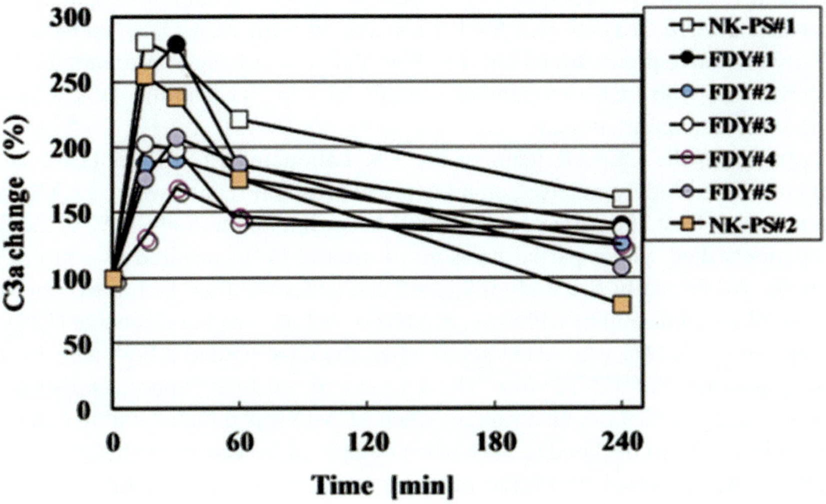

Fig. 6.5: Variation of C3a during the treatment time period of 4 hrs (119)

6.5 ECMO

ECMO is another vital medical device utilizing membranes as selective barriers to separate toxins from blood. In ECMO, the blood is pumped out of the patient's body into a membrane oxygenator, separating carbon dioxide (CO_2) and sending oxygen-rich blood back to tissues of the body. Blood is initially pumped from the heart's right side to the membrane oxygenator in the heart-lung machine, then rewarmed and returned to the body. The first ECMO treatment for adult respiratory failure was reported in the 1970s. The use of ECMO for respiratory patients has not significantly increased over the past 30 years. While ECMO treatment is well-recognized for neonatal respiratory failures, its acceptance among adult patients is gradually rising.

In case of respiratory failure, such as acute respiratory distress syndrome (ARDS), the lungs fail to supply adequate oxygen to the bloodstream, leading to organ failure and septic shock. The primary goal of treatment is to ensure the proper external oxygen exchange and sufficient oxygen delivery to the bloodstream. ECMO, a treatment mechanism that pumps oxygen and removes CO_2 is stripped out from it. In 1974, Dr. Robert Bartlett of Orange County Medical Center, USA, first used ECMO to treat a newborn baby named "Esperanza," whose arterial pressure had dropped to 12 mm Hg, leaving no survival chances. After three days of treatment,, the baby fully recovered. Dr. Bartlett then conducted several neonatal trials, resulting in a surprising 75% survival rate despite a predicted 90% mortality. However, ECMO testing on adult respiratory patients was uncommon at that time. The University of Michigan opened the first ECMO center in 1980. A 1989 comparative analysis at Boston's Children Hospital (USA) revealed a 90% survival rate for babies treated with ECMO, as compared to 67% with conventional treatment. In 1996, the United Kingdom reported a 70% survival rate with ECMO treatment, significantly higher than the 41% achieved with conventional treatment.

In 2009, the CESAR team in the UK conducted ECMO trials for adult patients. This study was based on a Glenfield Hospital report, where ECMO had been used to treat 200 patients since 1989. According to the study, 34% of the first 50 patients died with a partial pressure of arterial O_2 to inspired O_2 ratio of 65 mm Hg. Additionally, the study indicated that patients either died or experienced severe physical disability within six months or before hospital discharge (120). In spring 2009, ECMO was tested again when the CDC issued a high alert for the novel influenza A (H1N1) virus. The Extracorporeal Life Support Organization (ELSO), reported a 72% survival rate when ECMO was initiated within 6 days of intubation, but this declined to 31% when intubation exceeded seven days. ECMO is generally classified into three main types based on cannulation techniques: veno-arterial (VA), veno-venous (VV) and arterio-venous (AV).

6.5.1 Key Terms in Oxygen Transport Via Blood

Figure 6.6 shows the exchange mechanism of oxygen and carbon-dioxide (O_2/CO_2) within the blood. In the bronchial region, the oxygen pressure is maintained at 150 mm Hg. During the process, this pressure decreases to 100 mm Hg by the time it reaches the alveolar region. Moreover, within the blood vessel, the oxygen pressure of the deoxygenated blood is approximately 40 mm Hg, which is significantly lower than the oxygen pressure in the alveolar region. Consequently, due to this gradient in oxygen pressure, oxygen diffuses spontaneously from the alveolar region into the blood vessel, enriching the blood with sufficient oxygen.

Now, we will discuss how the diffusion process occurs within the alveolus and the problems associated with oxygen transfer. Once oxygen is transported to the blood, four oxygen molecules bind with hemoglobin, leading to 100% saturation of hemoglobin. However, if only three oxygen molecules bind, the result is 75% saturation of hemoglobin. Typically, the range of hemoglobin is 14–16 g/100 mL of blood for adult males, and around 12–15 g/100 mL for females. Each gram of hemoglobin can carry 1.34 to 1.39 mL of oxygen. Therefore, in adult males, the oxygen volume percentage in blood, considering hemoglobin as 100% saturated, will be (1.34 to 1.39)×(14 to 16) mL of oxygen per 100 mL of blood. Now, suppose the hemoglobin is 97% saturated the oxygen volume percentage in blood would then be (1.34 to 1.39)×(14 to 16)×0.97 mL of oxygen per mL of blood.

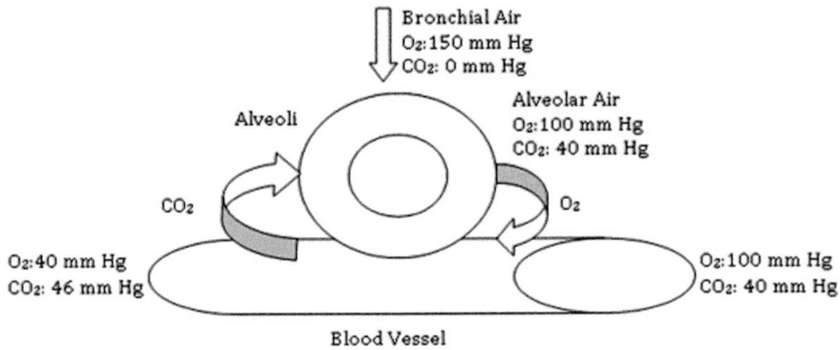

Fig. 6.6: Mechanism of exchange of oxygen (O_2) and carbon-dioxide (CO_2) in blood

It is well-known that at normal body temperature, the oxygen solubility in the blood is approximately 0.003 mL of oxygen/100 mL of blood for each 1 mm of Hg of partial pressure of oxygen (P_{O2}). So, for 100 mm of Hg $P_{O2,}$ the solubility of oxygen will be 0.3 mL of oxygen/100 mL of blood. Generally, only 2–3% of the oxygen remains in the dissolved state. During the oxygen transport mechanism, the total venous (Cv_{O2}), arterial (Ca_{O2}), and pulmonary capillary (Cc_{O2}) volumes of oxygen in 100 mL of blood are calculated using equations 6.3 to 6.5.

$$Cv_{O2} = \left[(1.34 \text{ to } 1.39)\times(14 \text{ to } 16)\times Sv_{O2}\right]+\left[Pv_{O2}\times 0.003\right] \quad (6.3)$$

$$Ca_{O2} = \left[(1.34 \text{ to } 1.39)\times(14 \text{ to } 16)\times Sa_{O2}\right]+\left[Pa_{O2}\times 0.003\right] \quad (6.4)$$

$$Cc_{O2} = \left[(1.34 \text{ to } 1.39)\times(14 \text{ to } 16)\times Sc_{O2}\right]+\left[Pc_{O2}\times 0.003\right] \quad (6.5)$$

In the above equations, the first part represents the volume of oxygen that will bind to hemoglobin, while the second part denotes the quantity of dissolved oxygen present in the bloodstream. There are also two additional terms: delivered oxygen (d_{O2}) and oxygen consumed by the tissues (V_{O2}), which fundamentally represent the oxygen uptake within the cells. When the cardiac output (CO), which is volume of blood pumped by the heart in one minute, the d_{O2} can be easily determined using this equation:

$$d_{O2} = CO\times Ca_{O2}\times 100 \text{ (mL of oxygen/minute)} \quad (6.6)$$

The normal CO value is 4900 mL of blood per minute. If the delivered oxygen volume is represented as the oxygen volume per 100 mL of blood per square meter of body surface area, it is referred to as a new term: 'cardiac index (CI).' 'CI' is essentially the ratio of cardiac output to the body surface area. Normally, the range of CI is 2.5-4 L of blood per minute per square meter of the human body's surface area. Equation 6.6 may also be represented as,

$$DI_{O2} = CI\times Ca_{O2}\times 10^5 \text{ (mL of oxygen per minute per square meter of the body surface area)} \quad (6.7)$$

6.6 Summary

- Discussion on dialysis and the principle of dialysis.
- Discussion on dialysis membranes and diffusion model.
- ECMO and the operating principle for ECMO.

CHAPTER

7

Micellar-enhanced Membrane Filtration

7.1 Introduction

In recent years, micellar-enhanced membrane separation has emerged as a promising alternative separation technique that requires mild operating conditions. In this process, the surfactant is initially added to the contaminated or polluted solution either at a concentration equal to or higher than critical micelle concentration (CMC). This eventually leads to the self-association of surfactant monomers to form micelles, which subsequently solubilize inorganic and organic pollutants within their interior. These micelles, containing the pollutants, are larger in size than the individual pollutants themselves, which facilitates their separation through a membrane. During membrane separation, the surfactant micelles aggregated with solubilized contaminants are retained by the appropriate membrane in the raffinate phase, while the free surfactant molecules and unsolubilized pollutants (in very minute concentrations) permeate across the membrane. Thus, with the application of this membrane-based strategy, treated water can be obtained, which can then be further recycled or discarded. Ultrafiltration (UF) membranes are generally preferred in this separation strategy over nanofiltration (NF) membranes. This is because UF membranes, having larger pore sizes, show higher efficacy in significantly retaining pollutants from wastewater and also provide greater permeate fluxes, offering economic benefits over NF and reverse osmosis (RO) membranes.

The primary concept of micellar-enhanced ultrafiltration (MEUF) was first explored in 1922 by J. McBain and W. Jenkins for ionic surfactants (potassium laurate and sodium oleate). Later, in 1964, H. Schott investigated the same process for the first time using a non-ionic surfactant (polyoxyethylated 1-dodecanol). Since its inception, numerous experimental investigations have been conducted on the application of MEUF technology in the field of wastewater treatment. Most of these MEUF studies have focused on the removal of heavy metals and dyes from aqueous solutions. The evaluation of the role of surfactant in pressure-driven membrane techniques has already been performed for metal retention, including copper, arsenic, cadmium, lead, nickel, manganese, zinc, boron, nickel, and cobalt. Existing literature has also explored the retention of several dyes, including mordant black 11, mordant black 17, and methylene blue. Some studies on MEUF have also reported on the treatment of wastewater in which the pollutants co-exist (25, 121-124). However, very few MEUF studies have been reported in the literature to treat real wastewater, including treatment of olive mill wastewater (125), phosphorus rich wastewater from the fertilizer industry (126), raisin processing wastewater, soft-drink processing wastewater (127), and plastic plating wastewater (128). Nevertheless, to the best of our knowledge, reports on the MEUF process for treating industrial dye wastewater are very rare.

The efficiency of the MEUF process largely depends on the type, characteristics, and concentrations of the surfactants and contaminants, wastewater composition, membrane pore size, ionic strength, pH, and hydrodynamic conditions including feed flow rate (v), transmembrane pressure difference (ΔP), stirrer speed (ω), temperature (T), etc. The primary limitation associated with the MEUF process is the transient reduction in permeation flux rate due to concentration polarization, pore obstruction, and membrane fouling. Concentration polarization refers to the reversible accumulation of solute particles in the liquid phase present just adjacent to the membrane surface. In the case of MEUF, it signifies the higher concentration of surfactant in the thin boundary layer adjacent to the membrane surface. However, membrane fouling occurs due to various factors, including solutes' (unsolubilized contaminants or small sized micelles') adsorption on the membrane pores, complete or partial blocking of the membrane pores, and the development of a deposited layer of retained contaminant-micelle aggregates near the membrane surface. During the progression of the concentration polarization phenomenon, surfactant micelles may induce complete or partial blockage of some membrane pores which eventually generates a resistance force against the solvent flow. As a result, a rapid reduction in permeate flux is observed at the initial stage of separation. During the MEUF process, gel layer formation may also occur due to the development of a highly viscous boundary layer by the retained pollutant-surfactant complexes near the membrane surface. When the concentration of surfactant micelles on the membrane surface becomes substantially high, the gel layer starts to form. Hence, this gel layer imposes additional resistance on top of the hydrodynamic membrane resistance, leading to a higher decline in permeate flux. Moreover, the extent of membrane fouling largely depends on the nature, characteristics, and size of the foulant.

7.2 Surfactant and Its Selection

A surfactant is generally considered a surface-active agent capable of changing the surface properties of a solution, leading to multiple effects and advantages. Introducing surfactants to any solution will alter its interfacial or surface tension by reducing the effective free surface energy of the interfaces. The surfactant adsorption is induced by the reduction in interfacial tension existing at the phase boundary. In this context, the degree of adsorption of surfactant molecules at the interface depends on the structure of the surfactant and the properties of the two phases that form the interface. A surfactant typically consists of two parts: a head and a tail. Generally, either a long hydrocarbon chain or aromatic ring along with a functional group acts as the head of the surfactant molecule. The tail of the surfactant is derived from unsaturated fatty acids of higher molecular weight or their derivatives. The presence of nucleophilic or electrophilic groups in the head of the surfactant accounts for its hydrophilicity (water-loving), while the tail part acts as the hydrophobic (water-hating) end of the surfactant molecule. The interaction of the hydrocarbon chain with water (H_2O) molecules in the aqueous phase is very poor. Conversely, the ionic or polar head group shows strong interaction with water molecules through ion-dipole or dipole interactions. Due to the strong interaction of the head with the H_2O molecules, the surfactant molecules become solubilized in water in a specific pattern, with the hydrophobic tail parts orienting themselves away from the aqueous phase. The symbiotic action of hydrogen bonding between water molecules and dispersion results in the squeezing of the long hydrocarbon chain outward, away from the water. Henceforth, these chains are referred to as hydrophobic. Any compound can be considered a surfactant only if it possesses the following three attributes: (i) the molecular structure of the compound should contain both polar and nonpolar groups; (ii) it should display surface activity, and (iii) it should develop self-assembled complexes or aggregates (such as vesicles, micelles, liquid crystalline) in the liquid phase.

Surfactants are amphiphilic in nature, as they contain both hydrophobic and hydrophilic groups. Commonly, surfactants are classified into four main types: anionic surfactants, cationic surfactants, non-ionic surfactants, and zwitter-ionic surfactants. Figure 7.1 represents the general classification of surfactants.

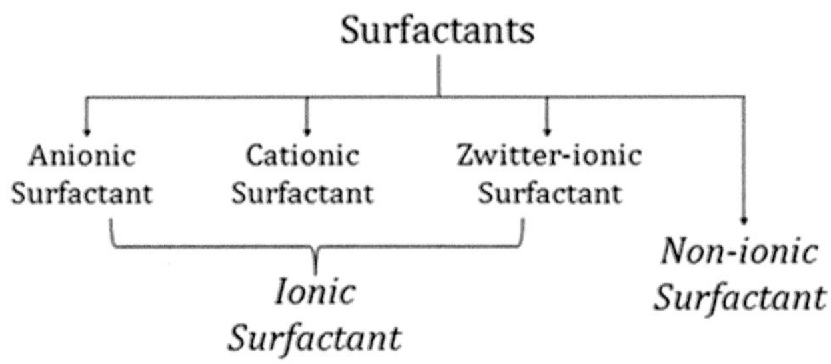

Fig. 7.1: General classification of surfactants

Among all of these categories, anionic surfactants are the most commonly and widely utilized ones due to the lower cost of manufacturing and wide domains of usage. These surfactants are negatively charged. Due to the presence of nucleophilic groups, when anionic surfactants are to aqueous solutions, negatively charged head groups are produced. Due to their negative charge, these types of surfactants exhibit high sensitivity to pH and electrolyte concentration. These anionic surfactants dissociate in water into monovalent, divalent, or trivalent metal ions. For instance, carboxylates dissociate into M^+ and $RCOO^-$, sulfates dissociate into M^+ and RSO_4^-, and sulfonates dissociate into M^+ and RSO_3^-, where M and R represent the inorganic or organic counter-ion and the hydrophobe, respectively. The hydrophobic chain generally consists of a linear arrangement of alkyl groups with 12 to 16 carbon atoms. Linear hydrophobic chains are generally preferred due to their higher effectiveness and easier degradability compared to branched ones.

Another category of surfactants is cationic surfactants (such as cetyl pyridinium chloride [CPC]). When dissolved in an aqueous phase, they generate positively charged heads due to the presence of electrophilic groups. These surfactants show high sensitivity to the electrolyte content and pH of the solution. Cationic surfactants exhibit a higher affinity for strong adsorption to solid surfaces. These surfactants may be formed by the direct or indirect attachment of a hydrophobic group to a protonated amino group, a quaternary ammonium group, or a heterocyclic base. Figure 7.2 is the general schematic representation of the structure of one of the commonly utilized cationic surfactants (quaternary ammonium compounds). Here, *R', R'',* and *R'''* denote the organic group while *X* denotes the halide group. The organic group may be aryl, alkyl, or alkyl-aryl group. These surfactants are water-soluble when there is only one long alkyl group present. Although these surfactants generally exhibit compatibility with most inorganic ions and hard water, they show incompatibility with metasilicates, highly condensed phosphates, protein-like compounds, and anionic surfactants.

Fig. 7.2: General schematic representation of the structure of a cationic surfactant (quaternary ammonium compounds)

The next class of surfactants is zwitterionic surfactants, which consist of both anionic and cationic groups. The properties of these surfactants are largely dependent on the solution pH. In the case of an acidic solution (pH below 7), the surfactant molecule will gain a positive charge. On the contrary, a surfactant molecule will gain a negative charge in alkaline solution (pH above 7). However, there is a specific pH value in which the surfactant has equal negative and positive groups, and this point is referred to as the isoelectric point. At this specific point, the surfactant exhibits both cationic and anionic nature, and the hydrophilic end becomes internally neutralized by the negative and positive counter-ions. Hence, the hydrophilic head of zwitterionic surfactants is usually larger in size because of their inherent ability for supporting both negative and positive charges, thereby exhibiting a higher affinity towards the water phase. These surfactants are usually introduced alongside anionic or non-ionic surfactants to modify properties such as solubility, micelle size, viscosity, foam stability, and detergency. The surface activity of these surfactants is largely influenced by the relative distance in between the opposite (negative and positive) charges, exhibiting the highest surface activity at their isoelectric point. The properties of this type of surfactants closely resemble those of non-ionic surfactants at the isoelectric point. Nevertheless, there are unique materials (such as sulfobetaines, sultaines, etc.) that are zwitterionic in nature at all pH values and, hence, do not possess any specific isoelectric point.

Another category of surfactant is the non-ionic surfactant, in which the electrical neutrality of the surfactant is beneficial, specifically because of the decreased tendency of interaction with other charged components in the solution. This characteristic of non-ionic surfactants facilitates their universal compatibility in any solution. The majority of these non-ionic surfactants are ethoxylated surfactants belonging from the polyoxyethylene family. Moreover, it has been interestingly observed that with an increase in solution temperature, surfactant solubility is reduced due to the inverse temperature relationship exhibited by non-ionic surfactants. The demerits of these surfactants include their availability primarily in paste or liquid form (rarely as solid pellets) and reduced solubility upon heating due to the inverse temperature relationship. Surfactants of the polyoxyethylene (POE) family can be represented by the general formula $RX(CH_2CH_2O)_nH$. Here, R represents the hydrophobic group (which can sometimes be a hydrophobic polyether e.g. polyoxypropylene), and X represents the nitrogen atom (N) or oxygen atom (O) or another functionality able to link the POE chain to the hydrophobe. Moreover, n signifies the total number of oxyethylene units present in the hydrophilic group. Generally, the properties of surfactants can be varied by altering the chain length of the hydrophobic group of POE, as in the case of n-dodecyl-POE surfactant.

7.2.1 Structure of the Surfactant

Any surfactant molecule typically has a hydrophilic part and a hydrophobic part, commonly referred to as the head and tail of the surfactant, respectively. The schematic representation of the structure of a surfactant is shown in Figure 7.3. The hydrophobic tail of the surfactant may be either linear or branched. Usually, one end of the alkyl chain is linked to the polar head group of the surfactant. The physicochemical attributes of the surfactant largely vary depending on key parameters such as the length, degree of chain branching, and the position of the polar groups. For instance, an increase in the length of the hydrophobic chain will result in the following events: (a) the solubility of surfactant in aqueous medium will decrease while increasing in organic solvents, (b) the density of packing of surfactant molecules at the interface will increase, (c) the tendency of adsorption of the surfactant molecules at an interface or the formation of micelles will be enhanced, (d) the melting point of surfactants, and (e) the probability of development of liquid crystal phases will also increase. If the counter-ions are present above a specific threshold concentration, there is a chance of precipitation of ionic surfactants in the aqueous medium. The relative sizes of the polar hydrophilic and hydrophobic groups, rather than the absolute sizes of any group, often play a decisive role in determining the physicochemical characteristics of a surfactant in the aqueous solution. Moreover, the introduction of unsaturation or branching into the hydrophobic tail group will result in: (a) enhanced solubility of the surfactant in aqueous medium or in organic solvents, (b) a reduction in the melting point of the surfactant and of the adsorbed film, (c) increased loose packing of the surfactant molecules at the interface (the 'cis' isomers are generally loosely packed while the 'trans' isomers are quite closely packed like the saturated isomer), (d) restricted formation of the liquid crystals phase in aqueous solution, (e) elevated oxidation tendency and formation of color in unsaturated compounds, (f) reduced biodegradability in branched-chain compounds, and (g) less thermal stability. Moreover, the introduction of an aromatic nucleus in the hydrophobic part increases the adsorptive behavior of the surfactant on polar surfaces, reduces its biodegradability, and results in molecules remaining loosely packed at the interface. The presence of polyoxypropylene units improves the hydrophobicity of the surfactant, its adsorptivity onto polar surfaces, and its solubility in organic solvents. However, the existence of polyoxyethylene units lowers the hydrophobic tendency of the surfactant. The presence of these above-mentioned groups in the hydrophobic part of the surfactant leads to a significant reduction in surface tension of water, much lower than that achieved with a surfactant containing a hydrocarbon-based hydrophobic part.

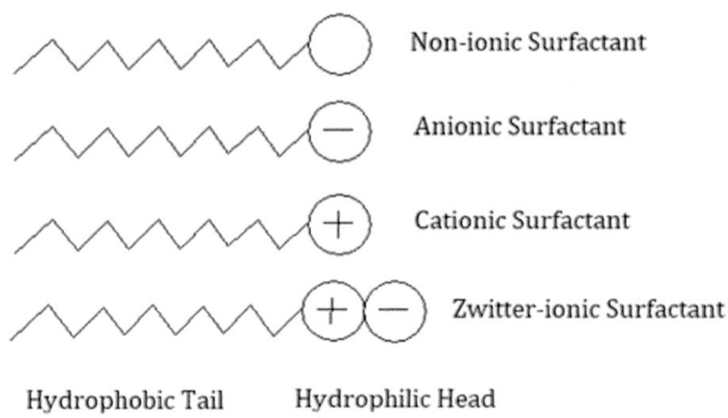

Fig. 7.3: Structural representation of different types of surfactants

7.2.2 Selection of Surfactant for Micellar-enhanced Filtration

To carry out the MEUF process, the utilization of surfactant is mandatory for promoting the formation of micelles, which are responsible to bind the contaminants or pollutants, and ultimately restrained by a suited membrane. Hence, formation of micelles and their characteristics are important factors in the entire filtration process. The binding of pollutants to the micelle based on electrostatic attraction is largely dependent on the size of the micelle, their distribution in the solution, and other parameters (such as temperature, pH, presence of co-ions, etc.). Thus, proper selection of surfactant requires significant attention to achieve high effectiveness in separating contaminants from wastewater. The extent of hydrophobic interactions at the water-micelle interface will significantly control the adsorption behavior of the solutes (pollutants), determining whether they will be adsorbed within the core of the micelle or at its surface. During the separation of metals and inorganic pollutants by the MEUF process, metal cations or inorganic pollutants attach to the oppositely charged head of the micelle's surface (see Figure 7.4) through electrostatic attraction. However, the mechanism differs in case of the removal of organic materials through MEUF. The organic pollutants usually get solubilized within the core of micelles (tail of the micelles) or palisade layer due to the Van der Waals force of attraction. Henceforth, taking into account the structure, size, and expense of the surfactant, it is essential to determine the minimum quantity of surfactant required for substantial adsorption of pollutants. Moreover, in this section, we will also address the effect of the molecular structure of a surfactant on the degree of solubilization of the solute.

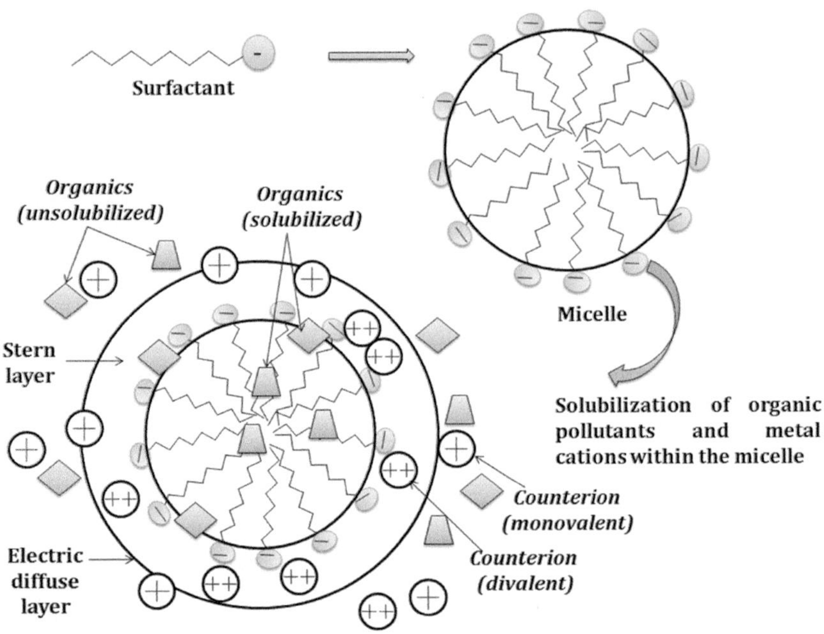

Fig. 7.4: Solubilization of organic pollutants and metal cations within the micelle

In the case of long chain hydrocarbons and polar compounds, which get solubilized within the interior or core of the micelle, the solubilization of contaminants usually increases with the size of the micelles. Thus, it can be inferred that any factor inducing an increase in either the micelle's diameter or its aggregation number is expected to facilitate the solubilization of the aforementioned type of material. Due to the increase in aggregation number with a higher degree of dissimilarity between surfactant and solvent, surfactants with longer hydrophobic chain lengths will exhibit higher solubilization capacity for hydrocarbon based contaminants in the core of the micelle within an aqueous phase. Nonionic surfactants exhibit superior solubilizing ability in very dilute solutions as compared to their ionic counterparts, due to their lower critical micelle concentration (CMC) values. Typically, the solubilizing capacity for long chain hydrocarbons and polar compounds within the inner core of micelles increases in the following order: Anionics<Cationics<Nonionics. The higher solubilization capacities of cationic surfactants compared to anionic surfactants with similar hydrophobic chain lengths may be due to the loose structural alignment or packing of the surfactant molecules in the formed micelles. Here are some appealing properties for selecting an ideal surfactant in MEUF:

- Higher solubilization capacity per unit weight of the dissolved surfactant.
- The critical micelle concentration (CMC) of the selected surfactant should be as low as possible.
- Facilitation of the development of large sized micelle structures should enable easy separation from the resulting solution through filtration, utilizing highly permeable membranes.
- Low monomer concentration to minimize surfactant wastage.
- Minimum tendency for phase separation problems (such as gelling, macroemulsion formation, precipitation, etc.).
- The Krafft point temperature of ionic surfactants should be very low. However, for nonionic surfactants, no specific temperature effect is observed on the property curve.

The long hydrocarbon chain of a surfactant is an attractive feature as it leads to larger sized micelles, higher solubilization capacity, and low concentration of monomer above CMC value. In this context, hydrocarbon chain lengths of anionic surfactants are confined to 12 carbons or fewer if the process is executed at ambient room temperature. For longer chain hydrophobic parts, the Krafft temperature exceeds room temperature leading to the precipitation of the surfactant and ultimately, the ineffectiveness of MEUF. However, micelles formed by nonionic surfactants exhibit higher solubilization ability per mole of surfactant and lower concentrations of residual monomers in aqueous solutions. Such surfactants are considered promising for utilization in the MEUF process. Though nonionic surfactants are often highly cost-intensive, rendering the MEUF process economically unviable, cationic surfactants of the same hydrophobic group size generally have lower Krafft temperatures compared to anionic surfactants.

Therefore, cationic surfactants with longer hydrophobic groups may be utilized in MEUF, leading to the formation of large micelles, higher solubilization capacities, and lower monomer concentrations. Moreover, the usage of cationic surfactants does not impose any major environmental risks. Cationic surfactants are typically the preferred choice for the MEUF process. Therefore, an ideal surfactant should have a low CMC value, a Krafft point above room temperature,the ability to form large sized micelles that can be completely rejected, and minimal or no solubilization of any unwanted solute.

7.3 Membrane Selection

Appropriate selection of a membrane is for achieving successful separation by micellar-enhanced membrane filtration. Key parameters to consider when assessing the membrane performance in the MEUF process include the material of the membrane, molecular weight cut-off (MWCO), and surface characteristics (such as charged, neutral, hydrophobic, or hydrophilic surfaces). The membrane material is selected based on the temperature, pH of the solution to be treated, and the solvent. For the MEUF process, polymeric membranes are predominantly used, synthesized from various materials, including polysulfone (PS), polyethersulfone (PES), polyamide (PA), polyacrylonitrile (PAN), polytetrafluoroethylene (PTFE), polyvinyldifluoride (PVDF), cellulose acetate (CA), and regenerated cellulose (RC). However, some case studies report the use of ceramic membranes in MEUF (129, 130). The advantage of ceramic membranes over polymeric ones lies in their sustainability under harsher operating conditions, such as highly acidic or basic conditions or elevated temperatures. One of the intricate issues in MEUF is membrane fouling. In small-scale laboratory experiments, membranes are intermittently cleaned between experimental runs to recover membrane performance by removing adsorbed solutes or surfactants. Nevertheless, after prolonged application of the membrane, fouling becomes intense, and cleaning may be insufficient. Moreover, membrane separation efficiency is often reduced due to biofouling, making membrane replacement essential. Previous literature reports that hydrophobic membranes exhibit lower fouling propensity, but permeate flux reduction due to surfactant adsorption is higher. In this context, the use of ceramic membranes may be a suitable alternative for addressing the fouling issue. The surface characteristics of the membrane significantly influence the adsorption of dissolved matter and surfactant, which, in turn, negatively affects the permeation phenomenon across the membrane. For instance, the study of Byhlin et al. (2003) demonstrated the application of the nonionic surfactant TX-100 in MEUF fitted with two different types of polymeric membranes: hydrophobic PES and hydrophilic RC membranes. The results showed that the adsorption of TX-100 is stronger on the surface of the hydrophobic PES membrane compared to the hydrophilic surface (131). Moreover, the surface characteristics of the membrane may also contribute to the lower retention of dissolved matters in the absence of surfactant. The MWCO of the membrane, which is correlated with the membrane pore size, must be selected thoughtfully based on the micelles size. Different polymeric UF membranes with varying MWCO in the range of 1–100 kDa are used in the MEUF process. However, in most of the cases, low MWCO (5 or 10 kDa) are utilized, as classical surfactants develop micelles with a hydrodynamic radius of 2–10 nm. A lower MWCO of the membrane may be preferred for MEUF systems containing smaller surfactant aggregates to achieve higher pollutant separation efficiency. However, membranes with lower MWCO generally exhibit lower permeate flux, leading to reduced efficiency for practical applications. The expense of the membrane can be reduced if high permeate flux is achieved along with the desired separation efficiency. Therefore, for MEUF systems with larger surfactant aggregates, the benefits of using membranes with higher MWCO may be utilized. The dependence of permeate flux on the MWCO of the membrane is sometimes relatively complex. In the case of pure solvents, the permeate flux increases with the MWCO as the membrane pore size increases. Though for micellar-based solutions the total permeate flux will also increase with the pore size, it does not follow the same pattern. Various other effects can occur, such as adsorption phenomenon within the membrane pores or on the surface, resulting in the development of a secondary resistance which will increase with higher MWCO. Besides the proper selection of the membrane, various procedures for pre-treatment and membrane cleaning should be mentioned. Prior to experimental runs, membranes are treated to condition them and eliminate any preservative products present. This is typically done by either soaking the membranes in distilled water or by filtering distilled water through the membrane. Intermittent cleaning of the membranes in between the experiments is conducted to remove adsorbed solutes and surfactants, thereby recovering membrane permeation. Various procedures for cleaning can be adopted. This may be done simply by washing with distilled water or by using a complex washing solution comprising sodium hydroxide and hydrochloric acid at very low concentrations. Irrespective of the cleaning process adopted, membrane permeability is evaluated prior to cleaning, and subsequently cleaned membranes are reused until the deviation from the initial water flux is less than 5–10%.

7.4 Mode of Operation and Membrane Module

These MEUF processes are primarily executed through two operational modes: dead-end mode and cross-flow mode. In the dead-end operational mode, the feed flow occurs in orthogonal to the membrane surface. After the rejection of the solute or surfactant by the membrane, their concentrations increase in the solution, resulting in a subsequent decline in flux over time. The primary reason for this is the gradual buildup of an adsorbed cake layer on the membrane surface, which increases in thickness over time. In this context, after reaching a certain solute concentration in solution, the permeation flux becomes almost zero, necessitating a pause in the filtration process. To improve permeation behavior, stirring of solutions is generally adopted. However, in the case of cross-flow mode, the tangential feed flow to the membrane occurs. Depending on the feed flow rate, a constant layer develops during steady-state operation, resulting in a constant permeate flux value. Therefore, compared to cross-flow, the feed flux rapidly declines over time in the dead-end operational mode due to the intense clogging of membrane pores. During cross-flow operation, a recycling system for permeate and retentate collection is used to avoid complete membrane obstruction and ensure continuous flow. This continuous flow system ensures the wide acceptability and applicability of cross-flow operation in industries, particularly when handling highly concentrated solutions. However, dead-end operation is typically adopted for laboratory scale MEUF experiments because of its easy execution, fewer equipment requirements, and lower volume utilization. Recently, in addition to the above-mentioned modes of operation, the MEUF process is also performed in a centrifuge system (132, 133). In centrifugal MEUF, the major driving potential is conveyed by the rotational speed of the centrifuge. In the near future, there is significant potential for upscaling and real industrial application of centrifugal MEUF.

The membrane configurations utilized for MEUF processes are flat sheet and hollow fibre configurations. However, in most of the existing MEUF studies, flat sheet membrane configurations were adopted. Flat sheet membranes are preferred for laboratory-scale experimentation due to the definite path of fluid flow and the acceptability of membranes with varied MWCO and high pressure ratings. Moreover, physical cleaning of the membrane can be easily carried out to mitigate intense fouling if required. On-site replacement of flat sheet membranes in the module is also comparatively hassle-free. On the contrary, the hollow fiber membrane configuration provides the largest surface area to volume (S/V) ratio and better permeation flux rate, and can be utilized for both cross-flow and continuous operations. The cleaning of such configurations is achieved through backflushing. The synergistic effect of higher flow rates and lower pressure drops results in reduced energy consumption, making hollow fiber configurations economical. Though very limited MEUF studies have been performed using hollow fiber membranes, these membranes have some limitations, including low pressure ratings, with a maximum TMP of less than 0.15 mPa. This restricts the flow rate of the feed stream in cross-flow mode. As the diameters of fibers are very small, they are highly susceptible to plugging at the cartridge inlet. Thus, if any single fibre within the bundle bursts the entire cartridge must be replaced, making the cost of membrane replacement very expensive.

7.5 Economical Aspect of MEUF

Regardless of the type of filtration process, the technical understanding of that process largely relies on its economic aspects. Being a highly complex process, the cost estimation of MEUF is dependent on various parameters. For a simplified cost evaluation, the expense can be fundamentally divided into two parts: fixed and variable costs. Fixed expenses will include the costs for the MEUF plant, while variable expenses will cover the costs associated with the MEUF process. In contrast to laboratory scale MEUF experiments, which consider small volumes and dead-end filtration modes, industrial MEUF plants treat bulk volumes of effluents, necessitating a continuously operating plant. In such scenarios, the total pumping cost, including the fixed costs for pumps and operating costs for energy is a major consideration. Some important decisive parameters for the total cost estimation are: (a) membrane flux rate, (b) membrane separation efficiency, (c) membrane stability and lifetime, and (d) complementary costs. A membrane with higher separation efficiency for both the pollutant and the surfactant will significantly reduce the need for an additional post-treatment step, thereby decreasing extra expenses. The permeation flux is also crucial as it determines the membrane area requirement. A higher flux rate is preferable for the MEUF process, but it is typically achieved using membranes with higher MWCO, lower surfactant concentrations or higher pressures. However, in all three scenarios, the membrane separation efficiency is affected. Thus, it is important to maintain optimized parametric conditions for the MEUF process. Another critical parameter is the stability and lifespan of the membrane, as the separation performance of membrane alters overtime due to the fouling phenomenon. In the case of irreversible fouling, membrane replacement incurs additional expenses each time. Unfortunately, very few cost estimation studies of MEUF systems are available in the literature. In contrast, cost estimation studies for conventional membrane-based separation processes for water treatment are widely available in the existing literature. The major difficulty is that cost analysis demands an in-depth insight into the system and is also a function of the wastewater that needs to be treated. Modern water treatment plants, excluding MEUF, can execute water treatment at an expense less than 0.3€/m^3 of water. If a MEUF wastewater treatment plant of the same capacity is assumed, the surfactant expense, the post-treatment expenses (including membrane replacement, recycling, discharge, etc.) and the costs for membrane cleaning and exchange are the deciding factors in the cost analysis. Even if the same costs for membrane exchange and cleaning are assumed for both processes, the surfactant cost needs to be additionally incorporated. However, more detailed investigations into cost estimation are essential.

7.6 Recovery and Regeneration of Surfactant from the Permeate and Retentate Stream

To make the MEUF process more cost effective and environmentally friendly, it is crucial to recover the surfactant from both the permeate and retentate streams before their final disposal to the environment. If the retentate phase of the MEUF system is considered, this stream is usually enriched with high concentrations of surfactants and pollutants (such as heavy metal ions, dyes, etc.) retained by the membrane. This can eventually contribute to secondary environmental pollution. To reduce the complexity associated with secondary pollution, the surfactants present in the concentrated retentate stream can be recovered and reused. For tackling this challenge, several techniques have been reported in the literature, including precipitation through the inclusion of counter-ions, alkaline treatment, acidification, solvent extraction, electrochemical treatment, foam fractionation method, reducing temperature below the Krafft point, and two-step chemical treatment, etc. In the case of the precipitation technique, the used surfactant may be precipitated by adding monovalent or multivalent counterions with an opposite charge to the used surfactant. For instance, if $CaCl_2$ salt is added at higher than its stoichiometric value, a significant quantity of anionic surfactant (SDS) will precipitate, resulting in the formation of calcium dodecyl sulfate (CDS). However, since CDS has less solubility in water, it cannot be reused directly in the same process. Therefore, CDS can be first converted to its sodium salt by adding Na_2CO_3 and then re-dissolved in water for reuse (134). In another strategy, 6N NaOH was initially added to the retentate stream containing metal ions (except Sr(II)) to precipitate the SDS surfactant. The precipitated SDS was then separated by centrifugation and reused in the MEUF process (135). In this method, precipitation of 45–55% SDS was achieved, and the efficiency of metal ion removal using the recycled SDS surfactant was approximately 50–58%. This demonstrates the feasibility and suitability of repeated SDS utilization. Among the methods mentioned, the acidification technique was found to be the simplest and most effective method for recovering used surfactant from the retentate stream of the MEUF process, showing promising outcomes in several previous studies. Approximately 58.1% recovery of SDS from the retentate stream was attained with this method by utilizing sulphuric acid (of pH 1.0). With the use of recovered SDS, 88.1% removal efficiency of Cd(II) metal ions was observed. Moreover, SDS may also be recovered through lowering the temperature of the surfactant solution to 4°C, which is below the Krafft point of the SDS surfactant (16°C).Additionally, the foam fractionation method has been utilized for recovering SDS and Cd^{2+} ions from the permeate stream of the MEUF process (136). Literature also reports the adoption of two step pre-treatment methods, such as acidification followed by ultrafiltration, and chelation followed by ultrafiltration, to remove Zn(II)/Cd(II) ions from the micelles of the SDS surfactant (137). In this method, H_2SO_4 was reported to be the best acidic agent and EDTA the best chelating agent. The optimized pH value was 4.4 for the successful reclamation of SDS, with 68.5% recovery of Zn(II) and 66.5% recovery of Cd(II). Furthermore, MEUF treatment using the reclaimed SDS could remove 89.6% of Zn(II) and 90.3% of Cd(II). The study affirmed that the recovered SDS, when using the two-step pretreatment method, performed more efficiently compared to the recovered SDS through the acidification method. In another surfactant recovery technique, the reclamation of cationic surfactant CPC from the permeate stream was carried out using a two-step chemical treatment (138). In the first stage, potassium iodide (KI) was mixed with the permeate stream containing surfactant CPC at a molar ratio of 1:1 KI to CPC. This eventually leads to the precipitation of cetyl pyridinium iodide (CPI). In the subsequent stage, CPI was transformed into a water-soluble chloride by adding cupric chloride ($CuCl_2$) to the precipitate at a ratio of 1:5 M CPI to $CuCl_2$. It was observed that around 90% recovery of CPC surfactant was possible by adopting this two-step chemical treatment technique. Another recovery method, based on the principle of alkaline treatment followed by subsequent acid precipitation and centrifugation was explored for the recovery of anionic biosurfactant Rhamnolipid (RHL) from the retentate stream of the MEUF process (139). Alkaline treatment of the retentate solution containing RHL, Cd(II), CV, and salt was carried out using aqueous NaOH solution at a pH of 9.0, resulting in approximately 99.5% recovery of cadmium ions as cadmium hydroxide. The subsequent acid precipitation followed by centrifugation leads to the recovery of RHL biosurfactant from the supernatant, which is nearly cadmium free. The maximum RHL recovery of 84 ± 1% was achieved under optimized conditions (pH: 2.4, centrifugation speed: 4755 rpm, time: 16.8 min). This study also reported that the rejection of CV at 97.6% remained entirely unaffected even after the reutilization of recovered RHL for three consecutive cycles. However, a slight reduction in the rejection of cadmium ions to 92% was observed. In an another study, the recovery of saponin surfactant from the permeate stream of the MEUF process was experimented with after the removal of MV from waste water by solvent extraction using two different solvents n-heptane and n-butanol (140). The results showed that n-butanol performed better than n-heptane for the recovery of saponin surfactant. For a specific saponin concentration of 350×10^{-3} kg.m^{-3}, approximately 85% recovery of saponin was achieved using a 1:1 volume ratio of n-butanol to aqueous reetha solution.

7.7 Challenges and Future Prospects of MEUF

As discussed in the previous sections, MEUF is a versatile membrane-based technique that can be readily used to remove various types of toxic contaminants from water and waste streams. Despite exhibiting very high removal efficacy, often quantitative in numerous examples, MEUF is still not considered a standard method for water treatment and management. There are a few publications that have demonstrated the utilization of the MEUF technique for real wastewater with comparable removal efficiencies to those attained in model systems. However, to the best of our knowledge, there is still no large or industrial scale water treatment plant adopting the MEUF process. The MEUF process is still associated with certain drawbacks that need to be overcome initially, along with technical realization.

The surfactant should have the capacity to bind toxic contaminants to separate them through filtration. However, as the surfactant is water-soluble, it is consequently considered a secondary contaminant. Moreover, as many experimented surfactants exhibit quite high CMC values and are not very prone to biodegradation, post-treatment is essential for the permeate stream containing monomeric surfactant. Fundamentally, a biodegradable surfactant with a lower CMC value and the ability to form large sized micelles for enhanced binding properties is required. Nevertheless, the number of available biosurfactants and research investigations with biosurfactants is still very limited. Moreover, the expenses involved in the synthesis of biosurfactants are comparatively higher than those of widely available synthetic surfactants.

Likewise the selection and application of a proper surfactant, it's easy recovery and reuse is another important aspect. As discussed in the previous section 7.5, surfactant recovery may be executed through several methods. However, sometimes these methods involve additional steps, such as secondary filtration, and often the utilization of auxiliaries for separating the contaminants from the formed micelles. Without the reuse of the utilized surfactant, the expense of the MEUF process will elevate, as new surfactant must be added for effective contaminant removal. This might be feasible only if the surfactant is quite cheap and there is no need for contaminant recycling. However, such MEUF processes will not be sustainable in nature.

One of the intricate problems in the MEUF process is the flux rate, which depends on the operating parameters (such as ΔP, T, etc.) and the composition of the aqueous solution/stream. In the case of a wastewater treatment plant (WTP), several cubic meters of wastewater must be treated in a short duration of time. For attaining a higher throughput, either the permeation flux rate or the available membrane area must be very high. Maintaining a very high membrane area is quite difficult due to the high separation costs involved. Therefore, when the membrane area is fixed, a higher and stable permeate flux is essential. However, the adsorption of surfactant molecules on the membrane surface, gel-layer formation, or concentration polarization significantly reduces the permeation flux in comparison to pure water. Based on the attributes, characteristics, and concentration of the surfactant, the flux rate may be enhanced by increasing pressure only until the "limiting" flux rate is attained. Furthermore, membrane fouling is a complex challenge in the MEUF process that restricts the long-term utilization of any ultrafiltration membrane. As discussed previously, the permeation flux may be enhanced with the aid of an external electric field, although this is quite a difficult task for larger-scale applications. Therefore, fouling resistant membranes enhanced with lower affinity towards surfactant adsorption are essential. Surface modification or tuning of membranes is a viable option in this context. As the flux is also dependent on the MWCO of the used membrane, usually a greater throughput may be achieved with the use of membranes having larger pores. However, in the case of the MEUF process, since the size of the formed micelles is often very small (in nm), lower MWCO membranes are preferred. In some instances, the surfactant is capable of generating large micelles. In such cases, larger MWCO membranes are preferable.

The expenses involved with water treatment utilizing MEUF, along with the environmental problems and characteristics of the available surfactants, may be the major root causes for making this membrane-based technique less favorable for large-scale operations. However, the fundamental principles of MEUF—separation of solutes, and selective binding—may be quite promising in fields beyond water treatment and management. One of such domain could be the separation and recovery of value added compounds including ingredients from biomass or dissolved noble metals. As already known, through the addition of ligands, the separation of metals from each other is possible. Moreover, the concentration of sugars or amino acids is also achievable from dissolved aqueous solutions. Another promising operational field for the MEUF process could be catalysis, specifically for the aspect of catalyst recycling (although very limited) research is available, the outcomes are noteworthy. Micellar solutions have gained significant attention in the field of catalysis as "green" solvents with several instances of successful catalysis observed (141, 142). Recently, micellar solutions have been studied for the sequential synthesis of pharmaceutical products (143). However, the reactions are still executed in batch mode. The application of MEUF focuses on developing a continuous mode of operation, allowing the catalyst to be recycled along with the retentate stream while the product is derived from the permeate solution. Favored partitioning of the product between the micelles and aqueous media is required to achieve successful application. Otherwise, the system needs to switch to a reverse MEUF process.

7.8 Summary

- Discussion on surfactant chemistry and the choice of surfactant for MEUF.
- Choice of membrane for MEUF.
- Discussion on the membrane modules and mode of membrane separation for MEUF.
- Understanding process economy for MEUF.

CHAPTER

8

Biosensors and Hydrogels

8.1 Introduction

One of the recent applications of membranes is in fabricating point-of-care devices, where the membrane is used as a purification tool to isolate the component of interest. Let's start with an example. Diabetes is one of the major diseases that India is predominantly suffering from. Measuring blood glucose with glucometer is now common practice. Typically, a drop of blood is placed on a strip, and the glucometer then displays the blood glucose level. The question arises: how does a glucometer detect only glucose, given the presence of other carbohydrates in the blood? The answer lies in two processes: an enzymatic reaction and the purification of blood glucose through a membrane that isolates glucose specifically. In short, a biosensor is a device that detects biological components or biomarkers. A hydrogel, on the other hand, is a water-filled material that may or may not surround a substrate. This substrate could be a drug or chemical used to detect biomarkers. This chapter will thoroughly discuss biosensors, their mechanisms of action with examples, and their mathematical principles. It will also cover hydrogels, including their preparation and applications in biomedical fields. Additionally, the chapter will provide an overview of the interplay between hydrogels and biosensors.

8.2 Biosensors and Membranes

In a broader sense, biosensor has three components which simultaneously act to devise efficient biosensor. The front part is the receptor, which is going to accept analytes to be measured, the second one is the transducer converting analogue signal into digital signal and the signal processing unit. Let us take one example of glucose level detection technology. Blood glucose (analyte) reacts with glucose oxidase (GOx) enzyme that produces gluconic acid and it changes the pH detected by an electrode followed by its conversion to a digital value. So the electrode here is the transducer that converts the pH into an electrical signal. In the chapter the primary question is how the blood reacts with the enzyme and is there any role of membrane here for the detection of glucose with high specificity. Now this specificity can be done through purification of analytes, which is done by semipermeable membrane like nafion, cellulose acetate, polypyrrole, polyurethane, chitosan and poly(2-hydroxyethyl methacrylate) (144). In glucose biosensor the blood sample comprises of several components if comes in contact with the working electrode may damage the electrode and the sensitivity of the biosensor will be reduced. Therefore, in between the blood sample and enzyme layer a semipermeable membrane is placed that selectively passes glucose towards the enzyme layer for further enzymatic reaction. Here in the biosensing devises enzyme is basically covalently linked with the membrane and may be called membrane bioreactor. But as discussed primarily one of the intricacies with the biosensing devices is the interference of other biomolecules in blood that may perturb the sensitivity of the measurement. In a literature (145) there are two membranes attached over the enzyme – one polydimethylsiloxane (PDMS) as a diffusion-limiting membrane on which enzyme is linked and at the cellulose acetate membrane to restrict the interference. Figure 8.1 shows the primary assembly of the setup, where CMOS is complementary metal oxide semiconductor (146). Although CMOS is related to the transduction process, till to understand the gross idea on the transduction a brief understanding may be needed. Now what is CMOS? To understand CMOS, let us first describe Metal Oxide Silicon Field Effect Transistors (MOSFET). It is a semiconductor device to switch or amplify voltage in the circuits. It has terminals source, gate, drain and body. Figure 8.2 is a schematic diagram for MOSFET. p-type semiconductor is heavily doped with n-type impurities and is shown as n+-type in the diagram. At the p-n junction there is a depletion zone as seen from the figure and is extending from source to drain as the gate voltage is increased. Now if the battery is getting connected with the gate and source, with positive terminal attached to the depletion layer thickness will be increased between source and drain. As one is increasing the voltage between source and gate, the thickness is continuously increasing with tapered region as shown by dashed line in the figure along with a current flow from drain to source. However, the details of the MOSFET are described somewhere else and out of the scope of this chapter. Whatever is described here is called N-channel MOSFET or in other way it is

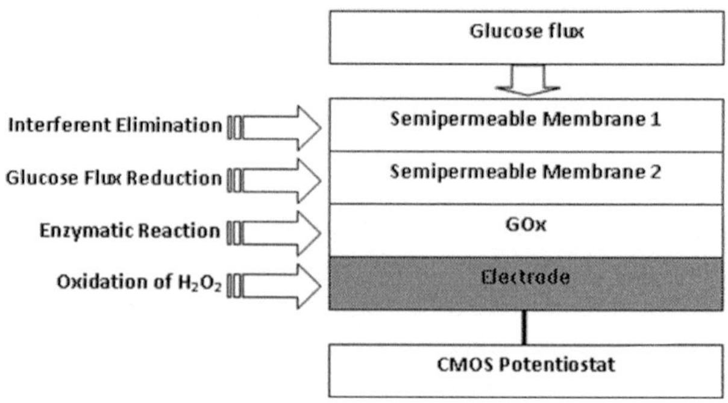

Fig. 8.1: Glucose sensing devise

called NMOS. Now if the substrate is n-type with p-type doping, then that is called P-channel MOSFET or PMOS. Combining these two is called CMOS (Figure 8.3).

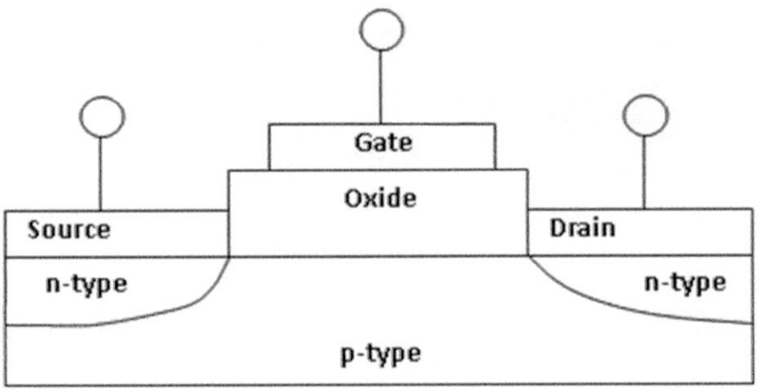

Fig. 8.2: Schematic of MOFSET

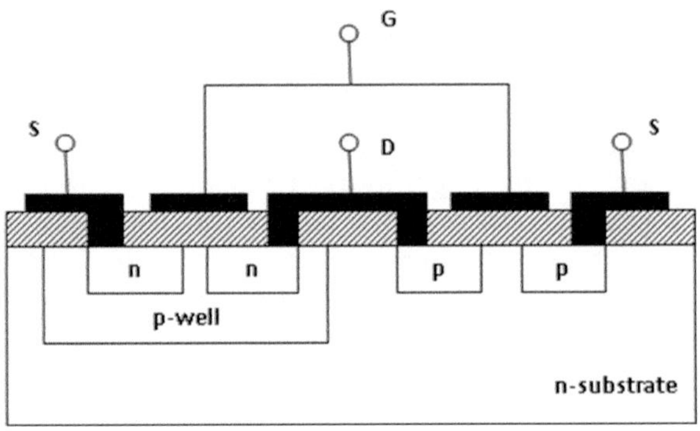

Fig. 8.3: Schematic of CMOS (S: Source; G: Gate; D: Drain)

It is evident that membranes play a significant role in devising biosensors in order to increase the specificity and purity of the analyte to be detected. Zhao and Ju (147) developed a multiwalled carbon nanotube (MWCNT) based membrane with GOx immobilized on it. Figure 8.4 (147) shows the schematic representation of fabricated MWCNT. A study utilized a gold electrode modified with a by-layer of poly(dimethyldiallylammonium chloride) (PDDA) and poly(sodium 4-styrenesulfonate) (PSS). The multiwalled carbon nanotube (MWCNT) was then wrapped with PDDA, and GOx was immobilized on its surface. Figure 8.4 (147) illustrates the schematic representation of the fabricated MWCNT. Five layers of the bilayer PDDA-MWCNT/GOx on modified electrodes demonstrated maximum analyte sensitivity. One key advantage of MWCNT is its membrane's electrocatalytic activity, which reduces oxygen (O_2) to hydrogen peroxide (H_2O_2) (equation 8.1)

$$O_2 + 2H^+ + 2e^- \rightarrow H_2O_2 \quad (8.1)$$

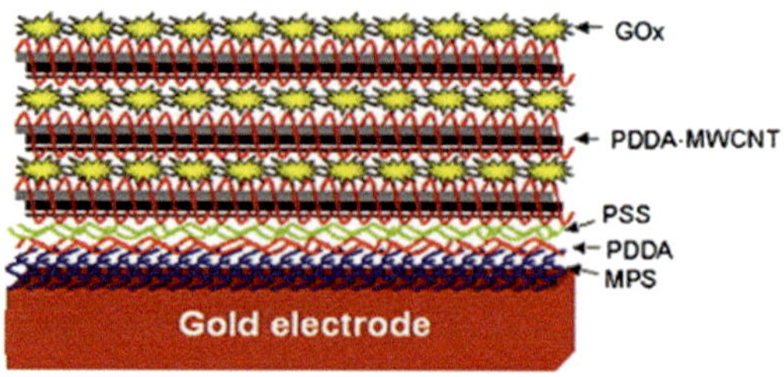

Fig. 8.4: Schematic of the GOx embedded MWCNT
(Reproduced with permission from Elsevier)

The description extends beyond blood glucose sensing and biomedical applications to include detecting food toxicants and environmental pollutants. Lipid membranes are among in biosensing, capable of detecting analytes like insecticides, pesticides, herbicides, metals, toxins, antibiotics, microorganisms, hormones, and dioxins. The fundamental concept in understanding membrane usage in biosensing is identifying key parameters that enhance membrane specificity and sensitivity to analytes. Nikoleli et al. (148) had developed a biosensor based on lipid membrane, where lipid film was coated on a graphene nanosheet dispersed on a copper wire. The membrane is a bilayer membrane fabricated through mixing Dipalmitoyl phosphatidylcholine (DPPC) and dipalmitoyl phosphatidic acid (DPPA). Finally this mixture was dispersed on a graphene coated copper wire mounted on a glass filter. Urease enzyme was incorporated into the bilayer lipid membrane (DPPA and DPPC). Figure 8.5 depicts this membrane, where filtered urea contacts the urease, into ammonium ion and carbon dioxide (CO_2). A primary issue with urea detectors is that protein contact may inactivate the sensor. Evolved CO_2 passes through the membrane to a detector, reducing to CO_2 CO_2^- (equation 8.2). Membrane permeation of CO_2 towards the detector ensures analyte purity.

$$CO_2 + e^- \rightarrow CO_2^- \qquad (8.2)$$

Instead of detailing various biosensor applications of biosensor, let's summarize the content with a mathematical approach to understand biosensor measurements.

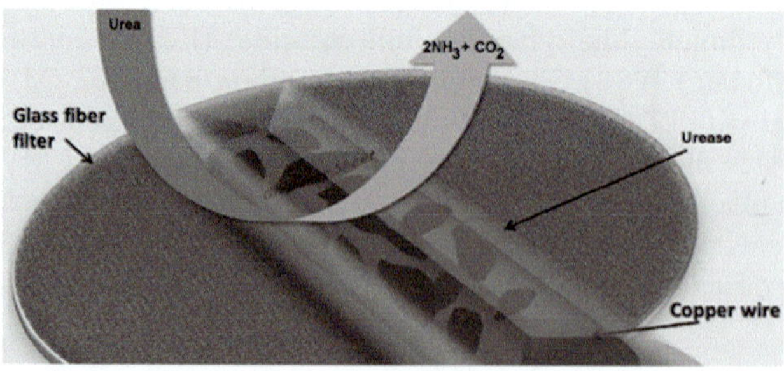

Fig. 8.5: Urea biosensor

8.3 Mathematics on Biosensor Membrane

Various mathematical models predict biosensor sensitivity, primarily based on the multilayer model with embedded enzyme, where analytes attach and release ions detected by the electrode. The mathematics of biosensors, including electrostatic diffusion models, and reaction kinetics are discussed below. The fundamental model for enzymatic biosensing devices is the Michaelis-Menten model. To understand the enzymatic model and the reactions that produce a signal through a transducer, let's start with some definitions. An enzyme is a biocatalyst (primarily proteins) that lowers the activation energy for a biochemical reaction by doing a specific job on a reactant (substrate). The basic enzymatic reaction, as described by Michaelis-Menten, is shown in equation 8.3.

$$E+S \underset{k_{-1}}{\overset{k_1}{\rightleftharpoons}} ES \xrightarrow{k_2} E+P \tag{8.3}$$

Inhibition, which reduces enzyme activity by binding with foreign material, reactants, or products, follows the non-Michaelis-Menten model and is given by equation 8.4.

$$\left.\begin{aligned} &ES+S \underset{k_{-3}}{\overset{k_3}{\rightleftharpoons}} ESS \xrightarrow{k_5} ES+P \text{ (substrate inhibition)} \\ &E+P \underset{k_{-4}}{\overset{k_4}{\rightleftharpoons}} EP \text{ (product inhibition)} \end{aligned}\right\} \tag{8.4}$$

E: Enzyme; S: Substrate; P: Product; ES: Enzyme substrate complex; ESS: Substrate inhibited enzyme; EP: Product inhibited enzyme.

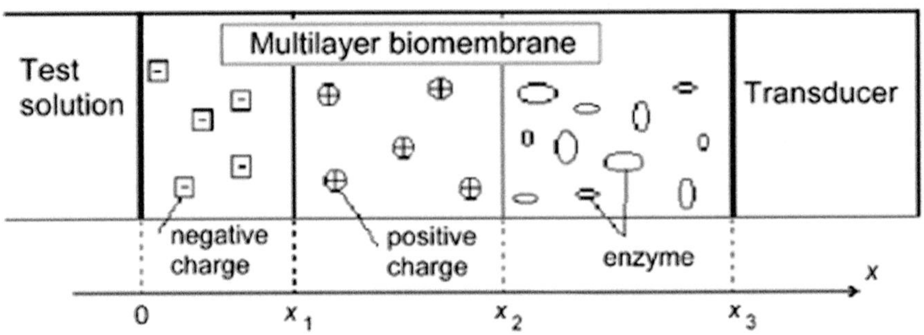

Fig. 8.6: Series model for biosensor (Reproduced with permission from Elsevier)

Focusing on the Michaelis-Menten model, Figure 8.6 presents a schematic diagram of a series attachment structure. Rossokhaty and Rossokhata (149) developed a biosensor model with three layers are considered. The first one for analytes, the next two for multilayer biomembranes with immobilized enzyme and the last for the transducer. The transient state electrostatic model is shown below.

$$\frac{\partial}{\partial x}\left(\varepsilon\frac{\partial\varphi}{\partial x}\right)=-\frac{4\pi}{\varepsilon_0}\sum_{i=1}^{n}ez_iC_i+Q \tag{8.5}$$

The transient species transport equation is given by,

$$\frac{\partial C_i}{\partial t}=-\frac{\partial}{\partial x}\left(C_i\frac{ez_iD_i}{k_BT}\frac{\partial\varphi}{\partial x}-D_i\frac{\partial C_i}{\partial x}\right)-U_i \tag{8.6}$$

Equation 8.3 can be redefined for glucose and oxygen substrates. A combined equation can be derived, assuming both follow the Michaelis-Menten equations.

$$\mathrm{GOx(FAD)+Glucose}\underset{k_{-1}}{\overset{k_1}{\rightleftharpoons}}\underbrace{\mathrm{Glucose-GOx(FAD)}}_{\mathrm{GEO}}\xrightarrow{k_2}\mathrm{Gluconic\ acid+GOx(FADH_2)} \tag{8.7}$$

$$\mathrm{GOx(FADH_2)+Oxygen}\underset{k_{-3}}{\overset{k_3}{\rightleftharpoons}}\underbrace{\mathrm{Oxygen-GOx(FADH_2)}}_{\mathrm{OEH}}\xrightarrow{k_4}\mathrm{Hydrogen\ Peroxide+GOx(FAD)} \tag{8.8}$$

GOx(FAD): Oxidised form of glucose oxidase enzyme flavin adenine dinucleotide (FAD) redox cofactor (EO); Glucose: G; Glucolactone: GA; Oxygen: O; Hydrogen peroxide: HPO and GOx($FADH_2$): Reduced form of glucose oxidase enzyme (EH).

The rate equation for both equations (8.7) and (8.8) can be written as,

$$\left.\begin{aligned}\frac{d[GEO]}{dt}&=k_1[G][EO]-(k_{-1}+k_2)[GEO]\\ \frac{d[OEH]}{dt}&=k_3[O][EH]-(k_{-3}+k_4)[OEH]\end{aligned}\right\} \tag{8.9}$$

Assuming a quasi-steady state for intermediate complexes, equation 8.9 can be simplified into an algebraic equation as follows:

$$\left.\begin{aligned}\frac{d[GEO]}{dt}&=0=k_1[G][EO]-(k_{-1}+k_2)[GEO]\\ \frac{d[OEH]}{dt}&=0=k_3[O][EH]-(k_{-3}+k_4)[OEH]\end{aligned}\right\} \tag{8.10}$$

Now, substituting [EO] = [EO^T] – [GEO], into the first equation of (8.10),

$$[GEO]=\frac{[G][EO^T]}{K_M^G+[G]};\ K_M^G=\frac{(k_{-1}+k_2)}{k_1} \tag{8.11}$$

where K_M^G is the Michaelis-Menten constant. Furthermore, the second equation is written with EH*, the initial concentration of the reduced enzyme form before the reaction given in equation (8.8).

At steady state with the initial stage of the second reaction, the total EH formed can be estimated as $EH^T = EO^T - EO = GEO$.

Similar to equation 8.11, the concentration of OEH can be written as,

$$[OEH] = \frac{[O][EH^T]}{K_M^O + [O]}; K_M^O = \frac{(k_{-3} + k_4)}{k_3}$$

$$\Rightarrow [OEH] = \frac{[O][G][EO^T]}{(K_M^O + [O])(K_M^G + [G])} = \frac{[O][G][EO^T]}{K_M^O[G] + K_M^G[O] + [G][O]}; K_M^G K_M^O \approx 0$$

order of $K_M^G \sim K_M^O \sim 10^{-2}–10^{-6}$ mol/l

(8.12)

Therefore,

$$\frac{d[HPO]}{dt} = k_4[OEH] = V_{\max}\frac{[O][G]}{K_M^O[G] + K_M^G[O] + [G][O]}; V_{\max} = k_4[EO^T]$$

(8.13)

Here, the concentration written as $[i]$ in equations 8.9 to 8.13 must be read as C_i further in the running text. Therefore, equation 8.13 will be re-written as,

$$\frac{dC_{HPO}}{dt} = -\frac{dC_O}{dt} = -U_O = k_4 C_{OEH} = V_{\max}\frac{C_O C_G}{K_M^O C_G + K_M^G C_O + C_G C_O};$$

$$V_{\max} = k_4 C_{EO^T} \qquad (8.14)$$

Therefore equation 8.6 can be reoriented for glucose after replacing U_G with equation 8.4 and considering U_G (Recombination rate) = $2U_O$ (Generation rate) for an efficient biosensor activity.

$$\left.\begin{aligned}\frac{\partial C_G}{\partial t} &= -\frac{\partial}{\partial x}\left(C_G\frac{ez_G D_G}{k_B T}\frac{\partial\varphi}{\partial x} - D_G\frac{\partial C_G}{\partial x}\right) + \left(V_{\max}\frac{C_O C_G}{K_M^O C_G + K_M^G C_O + C_G C_O}\right)\\ \frac{\partial C_O}{\partial t} &= -\frac{\partial}{\partial x}\left(C_O\frac{ez_O D_O}{k_B T}\frac{\partial\varphi}{\partial x} - D_G\frac{\partial C_O}{\partial x}\right) + \left(\frac{V_{\max}}{2}\frac{C_O C_G}{K_M^O C_G + K_M^G C_O + C_G C_O}\right)\end{aligned}\right\}$$

(8.15)

Solving equation 8.15 with appropriate boundary conditions determines the concentration of glucose and oxygen. However, the intricacy of the biosensor lies with the interference effect and might not provide a stable sensing performance. Moreover, enzyme complexity affects their ability to maintain a stable response to an analyte. Various materials can protect enzymes from environmental interference and support them in retaining their activity. The main idea is to develop a a technology that encapsulates the enzyme within a matrix, protecting it from environmental interference. Hydrogel can encapsulate enzymes, ensuring a controlled reaction with the substrate. But what exactly is hydrogel, and how does it interact with and protect enzymes? The next section elaborates on hydrogels and their applications in biosensing.

8.4 Hydrogel Biosensor

Hydrogel is a three dimensional structure cross-linked polymer structure capable of absorbing large amounts of water while maintaining its bonds, even when swollen. This occurs due to the presence of strong hydrophilic groups, which create a robust hydrogen bond network with water. The first application of hydrogel in contact lenses occurred in 1960, when Wichterle and Lim synthesized poly(2-hydroxethyl methacrylate) (PHEMA) for use in the contact lens industry. Hydrogel devices for biosensing can be formulated using two methods: immobilization and encapsulation. Before delving into the deep formulation strategies for both, let's first describe encapsulation and immobilization technologies.

Encapsulation is a strategy to encase the molecules, such as enzymes, within the polymeric matrix that functions like a membrane (Figure 8.7). In contrast, immobilization involves anchoring the enzyme onto a substrate (Figure 8.8). One of the most common forms of encapsulation technology is enzyme encapsulation within alginate, which forms a complex cross-linked structure with calcium chloride (Figure 8.7).

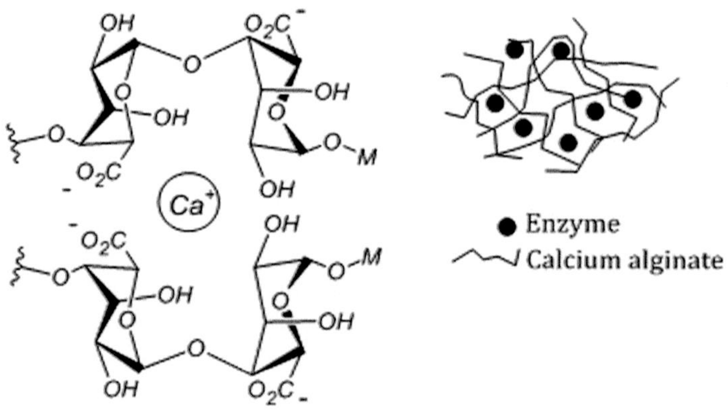

Fig. 8.7: Entrapment of enzyme

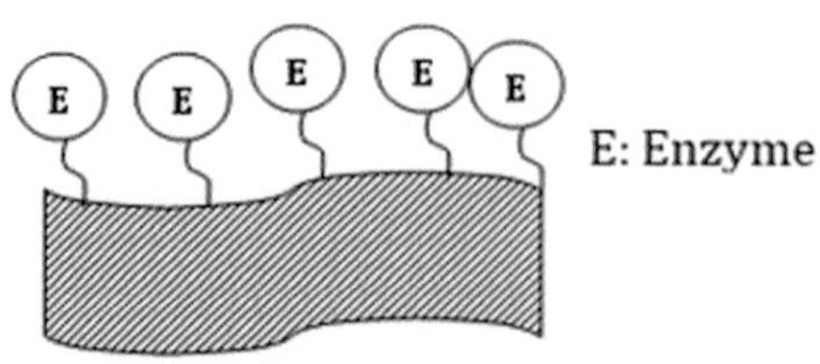

Fig. 8.8: Immobilization of enzyme on support

In a study made by Wang et al. (150) GOx was encapsulated within a sodium alginate/polyacrylamide hydrogel network. Additionally, zeolite imidazolate (ZIF-8)/calcium carbonate was used to further protect the encapsulated GOx. In contrast, Zhenglan et al. (151) immobilized GOx on a hydrogel optical fibre by mixing 2-hydroxy-2-methyl-propiophenone as a photo initiator, polyethylene glycol diacrylate (PEGDA) as a base, and deionized water in a specific ratio, followed by UV light irradiation. The first method detects concentrations in the range of 0.75–4 mg/mL,while the latter ranges of 20–100 µg/mL. The narrowing in detection range offers insight into the technology's performance mechanism. This is likely due to the higher specific surface area in encapsulation, resulting in better analyte sensing.

Biosensor applications extend beyond medical uses to include detection of organophosphorous pesticides (152) as point-of-care devices. The concept involves creating a color-inducing kit, called a target-responsive hydrogel-based kit, by embedding manganese oxide nanoflakes into sodium alginate integrated with 3,3′,5,5′-tetramethylbenzidine (TMB). Upon oxidation with manganese oxide, TMB turns blue, a change not achievable with Mn^{+2} ions. The oxidation of manganese oxide nanoflakes occurs in the presence of hydrogen peroxide, which is produced by the action of acetylcholinesterase (AChE) and choline oxidase (ChO) on acetylcholine (ACh). The schemes of the detection process are shown below.

Schemes

Scheme 1

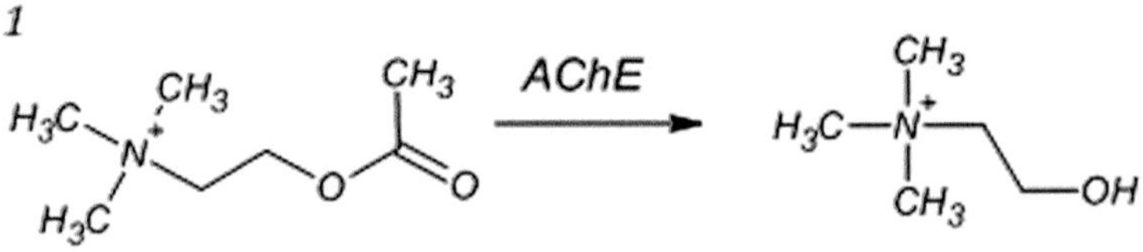

Scheme 2

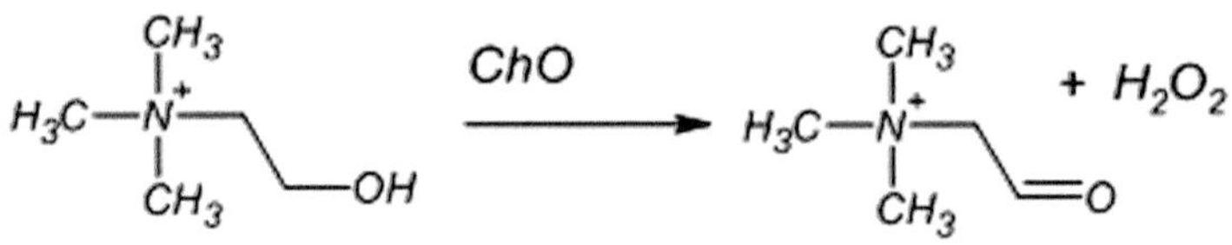

Scheme 3

$$H_2O_2 + MnO_2 + 2H^+ \longrightarrow 2H_2O + Mn^{+2} + O_2\uparrow$$

Inhibition of AChE suppresses hydrogen peroxide (H_2O_2) formation, resulting in less oxidation of manganese oxide nanoflakes into Mn^{+2} and consequently more oxidation of TMP. Increased presence of organophosphorous pesticide inhibits AChE further, reducing H_2O_2 generation, which in turn decreases TMP oxidation and reduces coloration. Another emerging technology involves preparing DNA hydrogel that recognizes the analyte through aptamer, triggering the release of encapsulated enzymes. One of the most commercially available biosensing devices is the personal glucose meter (PGM), where the glucoamylase enzyme is encapsulated. Upon contact with amylase, the aptamer facilitates the disintegration of the hydrogel, releasing the glucoamylase enzyme, which then converts amylose into glucose (153) for detection in the glucometer. A fundamental process for forming the 3D structure of DNA hydrogel involves pH variation to transform a linear motif into a 3D motif. To understand DNA replication, one must first grasp the Rolling-Circular Replication (RCR) process, depicted in Figure 8.9. Shahbazi et al. (154) detailed a strategy for formulating pure DNA hydrogel based on RCR technology. Though the strategy is complex and beyond this scope, a brief overview might spark interest.

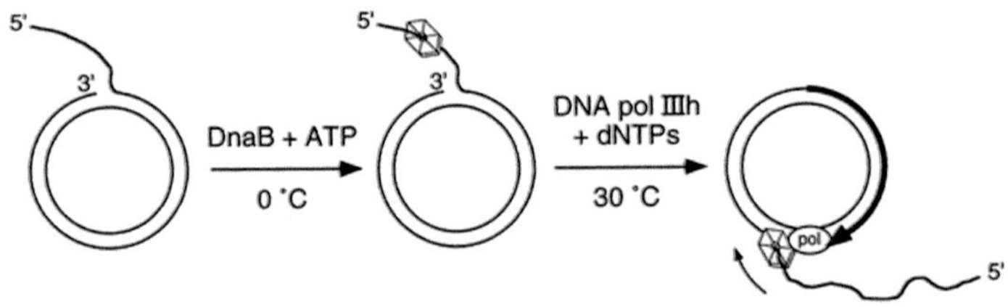

Fig. 8.9: Rolling-Circular Replication (RCR) process (DnaB: DnaB Heliase for opening up the replication fork during replication; ATP: Adenosine tri-phosphate; DNA pol IIIh: DNA polymerase to assemble nucleotides; dNTP: Deoxynucleotide triphosphate)

8.5 Mathematical Modeling of Hydrogel

Hydrogel modeling must consider swelling and the resulting convective motion imposed on its surface, known as the "swelling front." Setapa et al. (155) described this with a simple 1D analysis. Figure 8.10 shows the starting point at the centre, with the front travelling a distance $X(t)$.

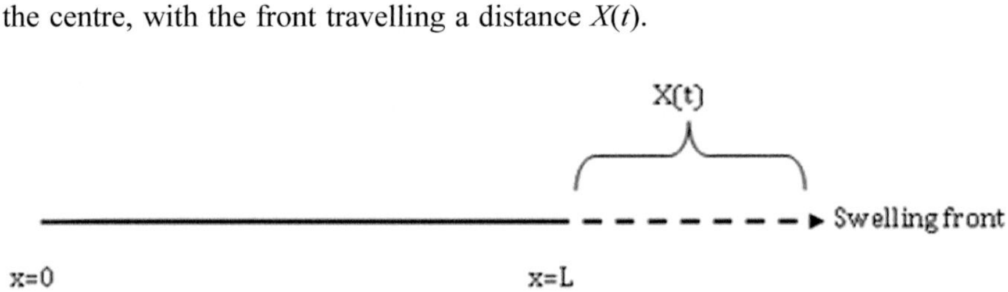

Fig. 8.10: Direction of swelling for hydrogel ($x = 0$ represents the centre of the hydrogel)

The mass balance equation can be written as

$$\frac{\partial c}{\partial t} = D\nabla^2 c - \nabla \cdot (cu) = D\frac{\partial^2 c}{\partial x^2} - c\frac{\partial u}{\partial x} - u\frac{\partial c}{\partial x} \tag{8.16}$$

Now $$\int_0^{X(t)} \frac{\partial u(x,t)}{\partial x} dx = \left.\frac{du(x,t)}{dx}\right|_0^{X(t)} = \frac{u(X(t),t) - u(0,t)}{X(t)} = \frac{\dot{X}(t)}{X(t)} \tag{8.17}$$

Therefore, following the same one can say,

$$u(x,t) = \frac{\dot{X}(t)}{X(t)} x \tag{8.18}$$

By substituting equations 8.17 and 8.18 into equation 8.16, it can be rewritten as,

$$\frac{\partial c}{\partial t} = D\frac{\partial^2 c}{\partial x^2} - c\left(\frac{\dot{X}(t)}{X(t)}\right) - \left(\frac{\dot{X}(t)}{X(t)} x\right)\frac{\partial c}{\partial x} \tag{8.19}$$

with the boundary conditions,

$$c(x, 0) = c_0; \quad \frac{\partial c(0,t)}{\partial x} = 0; \; c(X(t), t) = 0, \text{ where } 0 < x < X(t) \text{ and } t > 0$$

Introducing two new terms, ζ and τ, where $\zeta = \frac{x}{X(t)}; \tau = t$. Since ζ represents the ratio with the swelling front, the advection term becomes insignificant compared to the diffusion term, which becomes more prominent as the hydrogel size increases due to the additional space within the enclosure. Therefore, equation 8.19 becomes,

$$\frac{\partial c}{\partial t} = D\frac{\partial^2 c}{\partial x^2} - c\left(\frac{\dot{X}(t)}{X(t)}\right) \tag{8.20}$$

On replacement of the variables taken under consideration,

$$\frac{\partial c}{\partial \tau} = \left[\frac{D}{(X(t))^2}\right]\frac{\partial^2 c}{\partial \zeta^2} - c\left(\frac{\dot{X}(t)}{X(t)}\right) \tag{8.21}$$

The boundary conditions reappear as $c(\zeta, 0) = c_0$; $\frac{\partial c(0,\tau)}{\partial \zeta} = 0$; $c(1, \tau) = 0$, where $0 < \zeta < 1$ and $\tau > 0$. After transforming the moving domain, which now lies between 0 and 1 and is constant and time-independent, we apply the separation of variables technique,

$$c(\zeta, \tau) = A(\zeta)\, B(\tau) \tag{8.22}$$

Now replacing the variable in equation 8.21 and applying separation of variables technique,

$$A\frac{dB}{d\tau}=\left[\frac{D}{\left(X(t)\right)^2}\right]B\frac{d^2A}{d\zeta^2}-AB\left(\frac{\dot{X}(t)}{X(t)}\right)$$

$$\Rightarrow AB'=\left[\frac{D}{\left(X(t)\right)^2}\right]BA''-AB\left(\frac{\dot{X}(t)}{X(t)}\right)$$

$$\Rightarrow\left[\frac{\left(X(t)\right)^2}{D}\right]\frac{B'}{B}=\frac{A''}{A}-\left(\frac{X(t)\dot{X}(t)}{D}\right)$$

$$\Rightarrow\left[\frac{\left(X(t)\right)^2}{D}\right]\frac{B'}{B}+\left(\frac{X(t)\dot{X}(t)}{D}\right)=\frac{A''}{A}=-\lambda^2 \tag{8.23}$$

Therefore two simultaneous ODEs can be written as

$$\left.\begin{aligned}&B'+\left(\frac{\dot{X}(t)}{X(t)}+\frac{\lambda^2D}{\left(X(t)\right)^2}\right)B=0\\&A''+\lambda^2A=0\end{aligned}\right\} \tag{8.24}$$

Solving these ODEs,

$$\therefore A(\zeta)=a_1\cos(\lambda\zeta)+a_2\sin(\lambda\zeta) \tag{8.25}$$

and $A'(\zeta)=-a_1\lambda\sin(\lambda\zeta)+a_2\lambda\cos(\lambda\zeta)$

On applying boundary conditions,

$$a_2=0 \text{ and } a_1\cos(\lambda)=0 \text{ at } \zeta=1.\ a_1\neq0\ \therefore\cos(\lambda)=0\Rightarrow\lambda=\left(\frac{2n+1}{2}\right)\pi \tag{8.26}$$

Now with $B=c\exp\left(-\int_0^\tau\frac{\dot{X}(\tau)}{X(\tau)}d\tau\right)\exp\left(-\int_0^\tau\frac{\lambda^2D}{\left(X(\tau)\right)^2}d\tau\right)$

Again $\exp\left(-\int_0^\tau\frac{\dot{X}(\tau)}{X(\tau)}d\tau\right)=\left(\frac{1}{X(\tau)}-\frac{1}{L}\right)\approx\frac{L}{X(\tau)}\because L\gg X(\tau)$

$$B(\tau)=\frac{cL}{X(\tau)}\exp\left(-\int_0^\tau\frac{\lambda^2D}{\left(X(\tau)\right)^2}d\tau\right)$$

Therefore, $$c(\zeta,\tau)=\sum_{n=0}^{\infty}d_n\frac{L}{X(\tau)}\exp\left(-\left(\left(\frac{2n+1}{2}\right)\pi\right)^2\int_0^\tau\frac{D}{\left(X(\tau)\right)^2}d\tau\right)\cos\left(\left(\frac{2n+1}{2}\right)\pi\zeta\right) \tag{8.27}$$

After applying boundary condition $c(\zeta, 0) = c_0$, equation 8.27 gives,

$$d_n = c_0 \frac{4(-1)^n}{(2n+1)\pi}$$

Therefore, equation 8.27 can be rewritten as,

$$c(\zeta,\tau) = \frac{4L}{\pi} c_0 \sum_{n=0}^{\infty} \frac{(-1)^n}{X(\tau)(2n+1)} \exp\left(-\left(\left(\frac{2n+1}{2}\right)\pi\right)^2 \int_0^{\tau} \frac{D}{(X(\tau))^2} d\tau\right) \cos\left(\left(\frac{2n+1}{2}\right)\pi\zeta\right) \quad (8.28)$$

The main difference between hydrogels and conventional biosensors is the indirect interaction between the analyte and enzyme in hydrogels. This results in a reduced inhibitory effect on the enzyme, leading to lower noise and more sophisticated analyte detection.

8.6 Hydrogel: Scaffold Preparation

Hydrogel is increasingly being used in tissue engineering, particularly for scaffolds. Scaffolds are engineered biomaterials that support cells in tissue development. In living bodies, cells are usually supported by the extracellular matrix (ECM), which not only allows cell attachment but also provides mechanical strength to tissues and supplies growth factors. During tissue damage, materials that mimic the extracellular matrix (ECM) are essential. Hydrogels resemble ECM, can be processed under favorable conditions, and delivered in a minimally invasive manner. Therefore, materials for artificial cell support systems should closely resemble the natural environment. Hydrogel scaffolds can be broadly categorized into three types: space filling scaffolds, scaffolds for delivering bioactive materials and complex three dimensional networks that can support cell residence and subsequent tissue formation (156). Let's explore each type with a relevant case study.

8.6.1 Space Filling Materials

Pan et al. (157) developed a tunable and injectable hydrogel derived from ECM materials. The physicochemical and biological properties could be adjusted by varying the proportions of the base materials. They fabricated a gelable aldehyde-dextran (ald/dex)/amino gelatin (ami-gel) hydrogel by reacting sodium periodate with ethane diamine. The gelation density of the hydrogels could be adjusted by altering the proportion of (ald/dex)/(ami-gel). Crosslinking was achieved between amino group of the gelatin and the aldehyde group produced by oxidizing dextran with sodium periodate. Conversely, ethane diamine converts the carboxyl group on gelatin into amino groups. They estimated the swelling ratio, defined as the swollen hydrogel mass to dry hydrogel mass, by varying the ratios between dextran and gelatin. The swelling of the hydrogel ensures the release of encapsulated bioactive materials when the microenvironment becomes conducive for swelling. Their study showed that the maximum swelling ratio occurred with 30% dextran to 70% gelatin, while on the contrary the minimum was observed with equal ratios. Space-filling hydrogels provide a network for tissue regeneration damaged conditions. Once the tissue is built up, the hydrogel must degrade. The study found that the degradation rate was significantly lower with 70% dextran and 30% gelatin, which is undesirable. This suggests that dextran increases crosslinking, making the hydrogel harder and restricting the enzyme's permeation. Besides percolation through the hydrogel surface, cell adhesion and subsequent proliferation are crucial in evaluating the hydrogel's biocompatibility. A key consideration in using gelatin as a base material is that it is a partially hydrolyzed form of collagen, an active protein in ECM. The amino acid sequence is crucial for effective cell adhesion and proliferation. Effective cell culture was observed with a 50% dextran-gelatin combination. In another study by Liu et al. (158) icariin (IC) loaded modified halloysite nanotubes (mHCT) were incorporated into chitosan hydrogel to develop a nanocomposite hydrogel for bone tissue generation. However, detailed discussion on this topic is beyond the scope of this book. One of the primary points highlighted here is the assessment of hydrogels in space-filling scaffold fabrication. As mentioned, this type of scaffold requires proper porosity on the hydrogel surface, allowing pertinent biomaterials to percolate through and act as ECM. In the second example given, the crucial factor in repairing bone tissue is the process of generating blood vessels, known as "angiogenesis." Therefore, stem cells and various growth factors that are actually responsible for osteogenesis and angiogenesis are incorporated into the hydrogel structure. The critical aspect lies in the in vivo release of these substances, which is dependent on the porosity of the wall, controlled by the crosslinking (158).

8.6.2 Hydrogel Scaffold and Encapsulation for Delivering Biomaterials

Although this can be considered a new type of hydrogel scaffold, it is inevitable in scaffold engineering. As discussed earlier, hydrogel scaffold engineering for tissue regeneration of tissue must be assessed based on hydrogel swelling, crosslinking affecting porosity, and hydrogel degradation. Swelling is crucial for delivering materials encapsulated within the hydrogel, and this release depends on the hydrogel's swelling within the microenvironment. Let's elaborate on the hydrogel's swelling under the influence of pH. Equation 8.29 relates the pH of the microenvironment to the pKa (acid dissociation constant) value of the hydrogel's pendant group.

$$pKa = \mathrm{pH} - \log\frac{\left[A^{-}\right]}{\left[HA\right]} \tag{8.29}$$

where, HA is the pendant group. If the pH exceeds the pKa of the pendant group, the hydrogel's surface charge becomes more positive. Conversely, the other hydrogel displays a more negatively charged surface. In both scenarios water adheres to and percolates through the hydrogel network, causing it to swell. This swelling increases the hydrogel's porosity, leading to the release of bioactive materials. However, with the hydrogel scaffold, the primary condition is to ensure a slow degradation rate. Consider some examples cited by Sivaraj et al. (159), which mention that hydrogels are formed through a gelation mechanism, and the crosslinking occurs through covalent, ionic or physical bonds. Now, the cells are micrometer-sized, while the three dimensional mesh structure of the hydrogel is in the nanometer size range, leading to the entrapment of cells within the structure. When the hydrogel swells, nutrients, growth factors, ECM, and other components from the surrounding environment diffuse into it, aiding the cells in growing into tissue. A denser hydrogel mesh enhances cell-matrix interactions, making the hydrogel a supportive structure for cell growth in tissue.

Table 8.1: Hydrogel membranes and their applications

Material for hydrogel membranes	Application	Reference
Agarose/poly(acrylamide-co-acrylic acid) with varying acrylamide to acrylic acid ratio	Tissue replacements and soft robotics	(160)
15% (vol) Poly(ethylacrylate-*co*-methacrylic acid-*co*-1,4-butanediol diacrylate) (PEA-MAA-BDD) and microgels were swollen with an 85% (vol) acrylamide (AAm)/*N*,*N*′-methylenebisacrylamide (MBAAm) co-monomer solution	Tissue replacement	(161)
Polyethylene glycol diacrylate/gelatin methacrylate (PEGDA/GelMA) (cell encapsulation)	Treatment of bone tumors	(162)
basic fibroblast growth factor (bFGF)-collagen – silver sulfadiazine (AgSD) hydrogel	Treatment of burn wound inflammations	–
PVA/alginate hydrogel	Wound treatment of diabetic patients	
Alginate and methacrylated hyaluronic acid (Alg-HAMA) microfiber hydrogel to encapsulate pancreatic α and β cells	Controlling blood glucose and restricting the occurrences of hypoglycemia	(163)
Tannic acid grafted polyethyleneimine crosslinked polglycerate methacrylate hydrogel embedded polyether sulfone membrane (PES/G/E-TA)	Antioxidant	(164)

However, when using the scaffold as a delivery medium, the degradation of the hydrogel is a primary concern. Naturally, the key term that arises in this context is the rate of degradation. Let's consider a hypothetical scenario where cells are needed to heal the wound through tissue regeneration. The regeneration process will require time for the cells to multiply. If the degradation rate of the hydrogel is too high, it would hinder cell multiplication and, consequently, impede the wound healing process. Conversely, if the hydrogel degradation and the subsequent release are too slow, secreted factors may accumulate around the cell, negatively affecting the wound healing process. Thus, choosing the appropriate material and degree of cross-linking, which ultimately governs the meshing process, are crucial factors in determining the stability of the hydrogel and the release of encapsulated cells. Table 8.1 outlines various materials used for the preparing hydrogel membranes and their applications.

8.7 Summary

- Elaboration on the interrelation between biosensor and membrane.
- Elaboration on CMOS and MOFSET.
- Working principle of biosensors and biosensor model.
- Elaboration on hydrogel biosensors and their working principle.
- Hydrogel biosensor model development.
- Elaboration on hydrogel membranes and their applications.

CHAPTER

9

Membrane Separation Process Control

9.1 Introduction

The preceding chapters have discussed various applications of membrane and provided an overview of the variability in membrane technology related to different applications. However, in membrane technology no controlled applications have been designed yet, as membrane separations like MF, UF, NF and RO remain highly stochastic. It's challenging to determine when and how permeate flux decreases due to membrane fouling. Even in a controlled process, it is crucial to understand the error caused by disturbances, which can be mitigated by adjusting any of the process parameters. For instance, in a conventional membrane process, permeate flux is the measurable parameter, and fouling is the disturbance. In that scenario, to maintain the desired permeate flux, the engineer has two options: either open the control valve in the solvent feed line to dilute the medium or increase the trans-membrane pressure across the membrane. In the first scenario, adding diluents compromises product quality while in the second case the membrane might undergo damage through clogging or rupturing. Therefore, a binary approach cannot effectively address the issue with control applications on any membrane technology. Additionally, any deterministic process model may not be suitable due to the inherent qualitative and quantitative parameters associated with the membrane separation. To address the complexities in model development, many formulations now employ the concept of a probability distribution model. This approach aims to better reflect the real scenarios occurring during a membrane separation process. Moreover, probabilistic instances are more likely to occur with high shear membrane modules, necessitating greater attention to include these instances during formulation. In this context, Sarkar et al. (165) utilized the concept of surface renewal theory, incorporating two different distributions for time and space, to develop a predictive model for flux. However, the model's formulation becomes somewhat complex due to the usage of the distribution function, despite the remarkably high accuracy of the prediction. To bypass these complications, a new concept called the artificial neural network (ANN) intersects the field of mathematical formulation with membrane separation processes, which are highly nonlinear and involve numerous interactive parameters. Extensive research has already been conducted on modelling membrane separation processes using artificial neural networks (ANNs). This is largely due to their pattern mapping capability, which does not necessitate an in-depth understanding of the numerous interactive parameters involved in membrane-based separation processes. In 1995, Niemi et al. (166) proposed a simulation of reverse osmosis using artificial neural networks (ANNs). In 2006, Curcio et al. (167) conducted a similar study, modeling a cross-flow membrane separation process using a neural network-based approach. Bhattacharjee et al. (168) developed a predictive model using various neural network architectures to explore the advantages of artificial neural networks (ANN) in predicting permeate flux in a rotating disc membrane module (RDMM), a process that is complex to model using a mechanistic approach. On the other hand, artificial neural networks (ANN) offer the flexibility to extend the tool to highly nonlinear systems, based on the concept of mapping cause and effect. However, since ANN is a purely data-driven, process specific simulation tool, it is challenging to utilize it for tuning the operating parameters to optimally predict the system across a broad range without actual experimental observations. Therefore, a standalone ANN does not extrapolate the system's response to establish the model's applicability for a wide range of variations in the controlling parameters. In 2011, Sen et al. (169) integrated a conventional mechanistic model with the ANN, creating a knowledge-based hybrid neural network (KBHNN) to predict the flux in a cross-flow membrane system. Incorporating a mechanistic model with a black-box type ANN approach enhances the model's applicability across a range of operating conditions. Additionally, the KBHNN approach aids in evaluating process performance by understanding the interrelations between the controlling parameters. However, integrating ANN with a mechanistic model complicates the formulation strategy, much like the primitive mechanistic approach. To entirely avoid these assumptions and the narrow applicability of the model, incorporating fuzzy logic with ANN—known as the artificial neuro-fuzzy inference system (ANFIS)—has proven effective in addressing all the issues of previous approaches. One of the key benefits of ANFIS is its data-driven pattern mapping nature, similar to ANN, combined with the ability to elaborate on the qualitative aspects of the process through verbose descriptions. Recently, Sargolzaei et al. (170) developed a membrane flux predictive model using ANFIS. They found a strong agreement between the experimental results and the simulated outcomes, with a remarkably low percentage error of 1%. In this chapter, a case study is conducted to elaborate a predictive model using ANFIS, which evaluates the permeate flux from an RDMM equipped with vanes (171), followed by the application of a control system.

9.2 Theoretical Background

Before delving into the detailed architecture of ANFIS, let's first take a brief look at neural networks and fuzzy mathematics. The fundamental principle of neural networks is pattern mapping based on prior training and experience with experimental data input into the network. Figure 9.1 illustrates a basic neural network, where the feed layer initially accepts experimental information for training the network and generating outcomes. Subsequently, the same layer accepts random information, and the outcome is expected to be precise since the network has already been trained. The basic idea mimics the human neural system. When we touch a hot bowl, we immediately withdraw our hand. Our act as the input layer or acceptor. Once the sensation is processed through several neural systems (hidden layer), we feel the heat. The outcome is the action, i.e., withdrawing our hand. From the discussion, it's clear that this is a binary type of solution. Either we withdraw (possibly assigned as 1) or we don't (possibly assigned as 0). However, sometimes the situation doesn't reflect such a binary

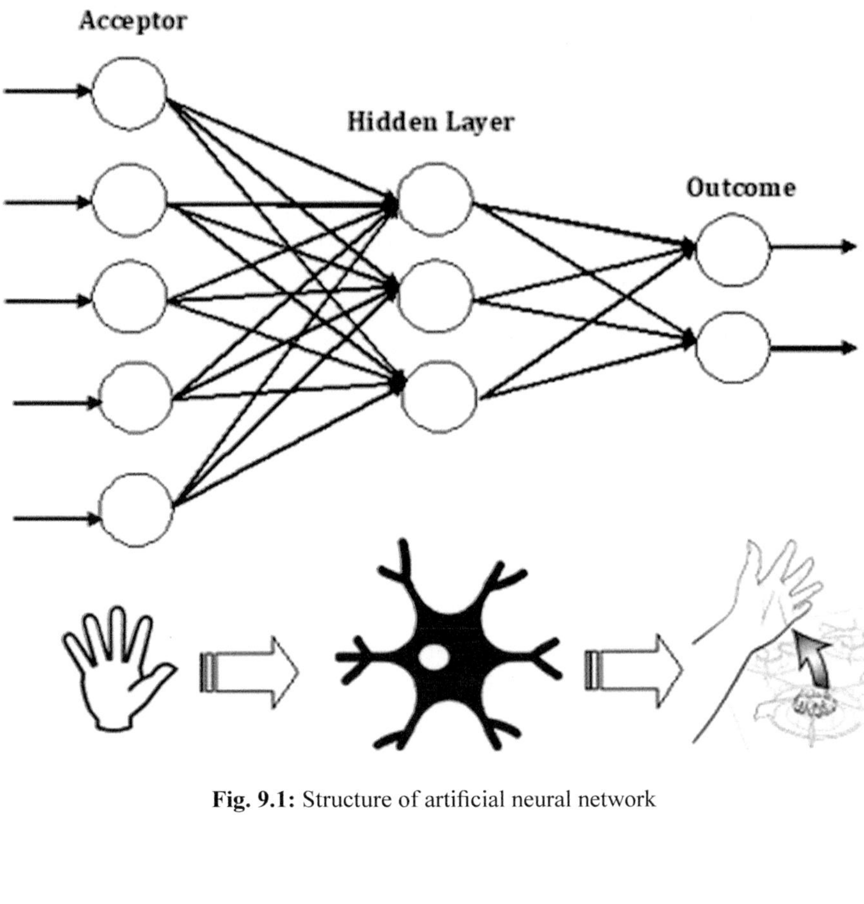

Fig. 9.1: Structure of artificial neural network

attitude. For instance, when addressing a person's eating habits to design a meal plan, the solution may not be strictly binary. A person may consume a moderate amount of food if they had a small meal a few hours ago. They might take only a light snack if they had a meal an hour ago. In the last scenario, the person may not eat anything if they had a meal just a minute ago. There are four possible scenarios, each with equal or varying probabilities. In this case, a binary approach will effectively design the network. They need to incorporate probabilities into the network, which is known as 'fuzzy' logic. To integrate this fuzzy mathematics into the neural network system, qualitative statements must be expressed mathematically to achieve measurable outcomes. This is precisely what ANFIS does, and it will be elaborated on next.

9.2.1 Architecture of ANFIS

In early 90's Jang (172) proposed a new simulation tool called ANFIS, which integrates the features of neural networks with fuzzy logic to enhance the applicability of neural networks from simple black-box simulations to grey-box simulations. There are two basic types of ANFIS networks: Mamdani and Sugeno. However, the suitability of a particular network for a system depends on the defuzzification procedure adopted for the individual networks. One of the critical issues in the defuzzification procedure for Mamdani is the center of gravity (COG) technique, which requires a thorough understanding of the process response to determine the member functions of the responses (Lancu (173). This becomes particularly critical when predicting responses in membrane separation processes. However, in a Sugeno type network, defuzzification is based on the weighted average of the inputs, thus eliminating the need for member functions for the responses. Therefore, adopting a Sugeno type network is advantageous in cases where the process is significantly multicollinear. Figure 9.2 illustrates a typical ANFIS architecture with Sugeno type linearization in the output. Similar to neural networks (Figure 9.1), ANFIS is a multilayer feed forward self-adaptive network that additionally integrates fuzzy logic interference with the progressive neural network layers. Layer 1 in Figure 9.2 essentially converts the qualitative understanding of the process into quantitative mathematical formulations using fuzzy-based "If-Then" rules. The selection of the function to quantify the qualitative concept using "If-Then" rules inherently depends on the dependencies of the process response (output from the network) with the operating parameters (input to the network).

9.2.2 Application of ANFIS to Predict Time Weighted Average Flux (TWAF) from RDMM Equipped with Turbulence Promoters or Vanes

The details of both ANFIS and ANN are elaborated in Table 9.1. As mentioned in the previous subsection 9.2.1., the complexity in ANFIS arises during the formulation of the qualitative rules that genuinely capture the essence of the

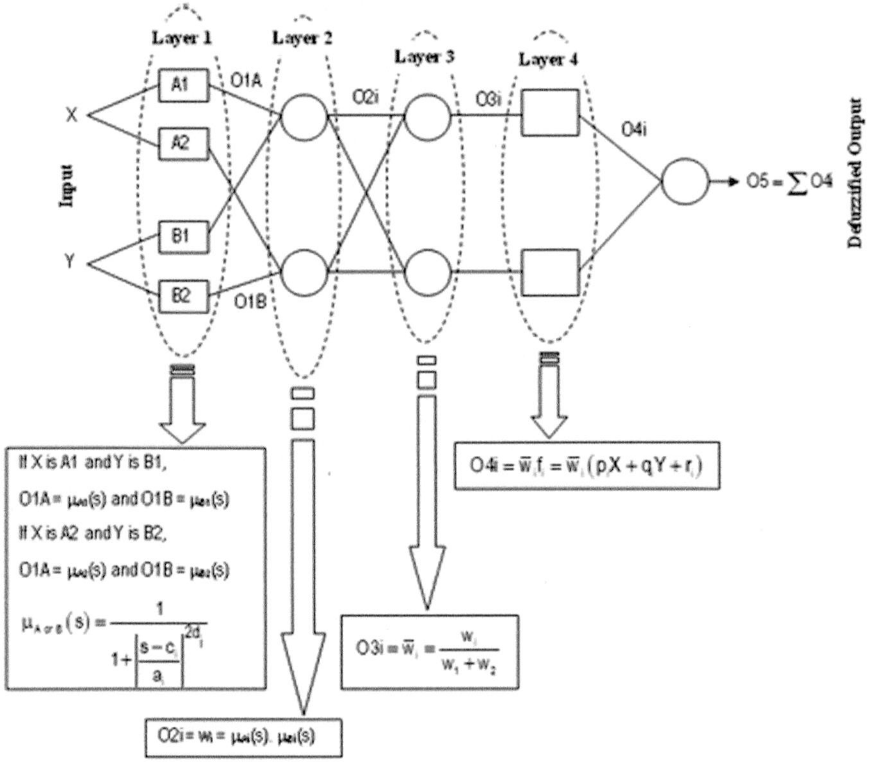

Fig. 9.2: Structure of an ANFIS network

experiment outcomes. In the current context, this complexity is heightened due to the inherent hydrodynamics associated with the vane geometries (Figure 9.3), where each vane type is detailed in Table 9.2. The values for time weighted average flux (TWAF) (equation (9.1)) are selected for different sets of membrane speeds and feed concentrations. All the values considered are at low TMP, concerning the reduced running cycle of the air compressor. According to the module's design (Figure 9.4), low TMP results in low air pressure over the membrane, which also ensures a reduced running cycle of the compressor. This is controlled by an on-off control valve (indicated as "CV" in the Figure 9.4) to store air within the air cylinder (indicated as "Air reservoir" in Figure 9.4) and maintain a certain TMP. However, the current network design benefits from the pattern mapping features of the ANFIS along with the set of rules detailed in Table 9.3. The verbose descriptions are based on the fundamental features of the vanes' geometry and the operability range. The ANFIS network was trained using squares back-propagation algorithm, while the ANN was trained using Levenberg-Marquardt back-propagation (LMBP) scheme. Another intricate issue in designing ANFIS is selecting a function that can convert the qualitative parameters into quantitative ones. To achieve these conversions from qualitative to quantitative, Gaussian functions were used, each corresponding to one of the vanes. These functions were discrete and non-overlapping, ensuring that the "predicted flux" for a given vane geometry would not affect the other five vane geometries.

$$\langle J \rangle = \frac{\sum_{k=0}^{final} J(t) t_k}{\sum_{k=0}^{final} t_k} \tag{9.1}$$

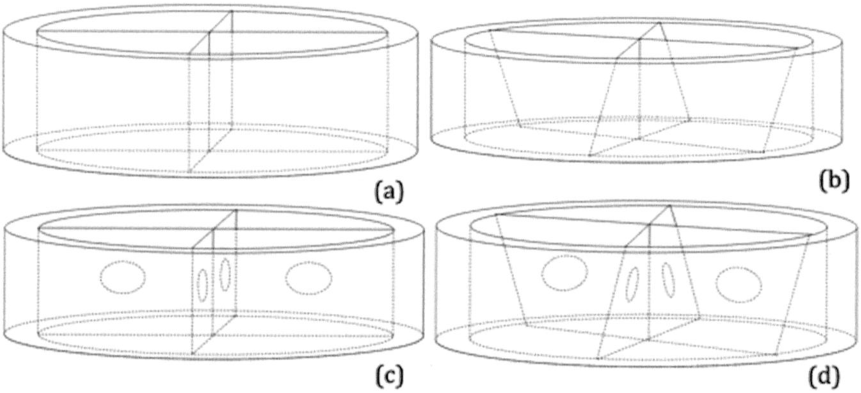

Fig. 9.3: Different types of vane geometries: (a) 90° blade angled vane fixed or rotating with the membrane; (b) 45° blade angled vane fixed or rotating with the membrane; (c) 90° blade angled vane with holes on the blade surface rotating with the membrane; (d) 45° blade angled vane with holes on the blade surface rotating with the membrane

In both ANFIS and ANN, 60% of the data set was used for training, while the remaining 40% was designated as the validation data set. The efficacy of both the networks was evaluated based on accuracy, measured by the percentage absolute average deviation (AAD) (equation (9.2)) between the observed and validated predicted results, along with the process time. ANFIS showed an AAD of 2.9% after 20 iterations, whereas ANN exhibited an AAD of 28.4% for the same validation data chosen for ANFIS, requiring 1297 iterations to converge, indicating a longer process time compared to ANFIS.

$$\text{Percentage AAD} = \frac{\sum_{k=1}^{N} \frac{\left| \langle J \rangle_k^{\text{Exp}} - \langle J \rangle_k^{\text{Pred}} \right|}{\langle J \rangle_k^{\text{Exp}}} \times 100}{N} \tag{9.2}$$

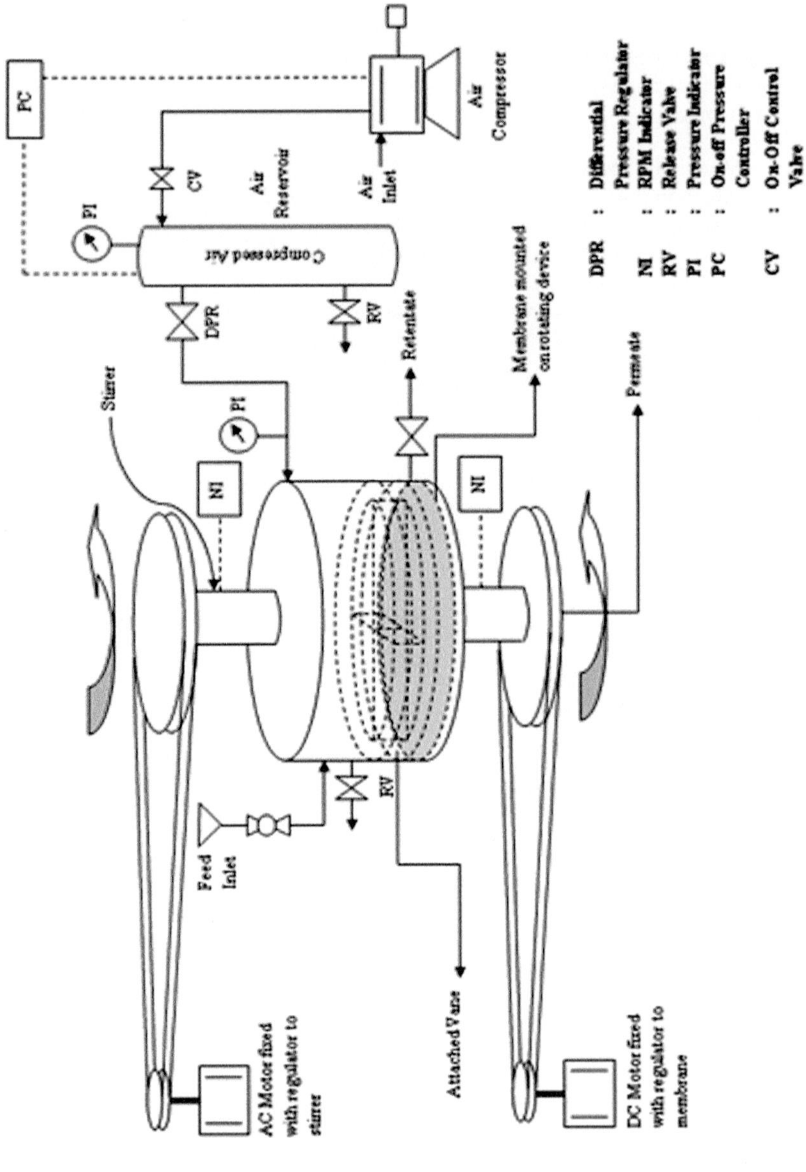

Fig. 9.4: Schematic diagram of RDMM (Reproduced after permission from Elsevier)

Table 9.1: Design details of ANN and ANFIS network for predicting TWAF (ms^{-1}) from high sheared membrane module RDMM

Network type	Input parameters	Output parameters	Learning algorithm
ANN	Membrane speed (rad s^{-1}), Feed concentration (kg m^{-3}), Vane number	TWAF (m s^{-1})	LMBP
ANFIS	Membrane speed (rad s^{-1}) *Function used*: Gaussian Lower range: >0.4*-δ10.5 Upper range: >10.5-δ20.9 Moderate range: >20.9-δ41.9 Feed concentration (kg m^{-3}) *Function used*: Gaussian Lower range: 0-δ0.05 Upper range: >0.05-δ0.1 Moderate range: >0.1-δ0.3 Blade geometry *Function used*: Triangular Range Vane angle 90°: Peak at 1 Vane angle 45°: Peak at 2 Vane angle 90° with holes: Peak at 3 Vane angle 45° with holes: Peak at 4 Vane Dynamics** *Function used*: Gaussian Fixed: 0-δ0.4* Rotating with, Lower range: >0.4-δ10.5 Upper range: >10.5-δ20.9 Moderate range: >20.9-δ41.9	TWAF (m s^{-1}) *Function used*: Gaussian Higher than high: ε12.5-<15 High: ε10-<12.5 Upper moderate: ε7.5-<10 Lower moderate: ε5-<7.5 Low: ε2-<5 Lower than low: ε0-<2	Least square back-propagation algorithm

* Based on the module's design, the membrane cannot be rotated at speeds below 0.5 rad s^{-1}. Any value below this threshold results in a static membrane.

** As the vanes rotate with the membrane, the membrane speed is considered as the domain of the function for the vane speed.

Table 9.2: Description of the vanes attached to RDMM over the membrane

Vane type	Vane description
1	Vane with a 90° blade angle and fixed relative to the membrane
2	Vane with a 45° blade angle and fixed relative to the membrane
3	Vane with a 90° blade angle and fixed relative to the membrane. The blades have holes at the surface.
4	Vane with a 45° blade angle and fixed relative to the membrane. The blades are having holes at the surface.
5	Vane with blade angle 90° and rotating with the membrane.
6	Vane with blade angle 45° and rotating with the membrane.

Table 9.3: Verbose in setting up the rule for designing ANFIS

Rule number	Rules
1	If (the vane angle is 90°) and (the vane is fixed with respect to the membrane) and (the membrane speed is low) and (the bulk feed concentration is low) then (TWAF is higher than high).
2	If (the vane angle is 90°) and (the vane is fixed with respect to the membrane) and (the membrane speed is high) and (the bulk feed concentration is low) then (TWAF is high).
3	If (the vane angle is 90°) and (the vane is fixed with respect to the membrane) and (the membrane speed is low) and (the bulk feed concentration is high) then (TWAF is lower moderate).
4	If (the vane angle is 90°) and (the vane is fixed with respect to the membrane) and (the membrane speed is high) and (the bulk feed concentration is high) then (TWAF is upper moderate).
5	If (the vane angle is 90°) and (the vane is fixed with respect to the membrane) and (the membrane speed is moderate) and (the bulk feed concentration is moderate) then (TWAF is lower moderate).
6	If (the vane angle is 45°) and (the vane is fixed with respect to the membrane) and (the membrane speed is low) and (the bulk feed concentration is low) then (TWAF is upper moderate).
7	If (the vane angle is 45°) and (the vane is fixed with respect to the membrane) and (the membrane speed is high) and (the bulk feed concentration is low) then (TWAF is upper moderate).
8	If (the vane angle is 45°) and (the vane is fixed with respect to the membrane) and (the membrane speed is low) and (the bulk feed concentration is high) then (TWAF is upper moderate).
9	If (the vane angle is 45°) and (the vane is fixed with respect to the membrane) and (the membrane speed is high) and (the bulk feed concentration is high) then (TWAF is upper moderate).

Contd...

Contd...

Rule number	Rules
10	If (the vane angle is 45°) and (the vane is fixed with respect to the membrane) and (the membrane speed is moderate) and (the bulk feed concentration is moderate) then (TWAF is upper moderate).
11	If (the vane angle is 90° with holes on blade) and (the vane is rotating with the membrane) and (the membrane speed is low) and (the bulk feed concentration is low) then (TWAF is high).
12	If (the vane angle is 90° with holes on blade) and (the vane is rotating with the membrane) and (the membrane speed is high) and (the bulk feed concentration is low) then (TWAF is lower moderate).
13	If (the vane angle is 90° with holes on blade) and (the vane is rotating with the membrane) and (the membrane speed is low) and (the bulk feed concentration is high) then (TWAF is high).
14	If (the vane angle is 90° with holes on blade) and (the vane is rotating with the membrane) and (the membrane speed is high) and (the bulk feed concentration is high) then (TWAF is lower moderate).
15	If (the vane angle is 90° with holes on blade) and (the vane is rotating with the membrane) and (the membrane speed is moderate) and (the bulk feed concentration is moderate) then (TWAF is high).
16	If (the vane angle is 45° with holes on blade) and (the vane is rotating with the membrane) and (the membrane speed is low) and (the bulk feed concentration is low) then (TWAF is lower moderate).
17	If (the vane angle is 45° with holes on blade) and (the vane is rotating with the membrane) and (the membrane speed is high) and (the bulk feed concentration is low) then (TWAF is lower moderate).
18	If (the vane angle is 45° with holes on blade) and (the vane is rotating with the membrane) and (the membrane speed is low) and (the bulk feed concentration is high) then (TWAF is upper moderate).
19	If (the vane angle is 45° with holes on blade) and (the vane is rotating with the membrane) and (the membrane speed is high) and (the bulk feed concentration is high) then (TWAF is lower moderate).
20	If (the vane angle is 45° with holes on blade) and (the vane is rotating with the membrane) and (the membrane speed is moderate) and (the bulk feed concentration is moderate) then (TWAF is upper moderate).
21	If (the vane angle is 90°) and (the vane is rotating with the membrane) and (the membrane speed is low) and (the bulk feed concentration is low) then (TWAF is low).
22	If (the vane angle is 90°) and (the vane is rotating with the membrane) and (the membrane speed is high) and (the bulk feed concentration is low) then (TWAF is lower moderate).

Contd...

Contd...

Rule number	Rules
23	If (the vane angle is 90°) and (the vane is rotating with the membrane) and (the membrane speed is low) and (the bulk feed concentration is high) then (TWAF is upper moderate).
24	If (the vane angle is 90°) and (the vane is rotating with the membrane) and (the membrane speed is high) and (the bulk feed concentration is high) then (TWAF is upper moderate).
25	If (the vane angle is 90°) and (the vane is rotating with the membrane) and (the membrane speed is moderate) and (the bulk feed concentration is moderate) then (TWAF is upper moderate).
26	If (the vane angle is 45°) and (the vane is rotating with the membrane) and (the membrane speed is low) and (the bulk feed concentration is low) then (TWAF is low).
27	If (the vane angle is 45°) and (the vane is rotating with the membrane) and (the membrane speed is high) and (the bulk feed concentration is low) then (TWAF is lower moderate).
28	If (the vane angle is 45°) and (the vane is rotating with the membrane) and (the membrane speed is low) and (the bulk feed concentration is high) then (TWAF is lower moderate).
29	If (the vane angle is 45°) and (the vane is rotating with the membrane) and (the membrane speed is high) and (the bulk feed concentration is high) then (TWAF is lower moderate).
30	If (the vane angle is 45°) and (the vane is rotating with the membrane) and (the membrane speed is moderate) and (the bulk feed concentration is moderate) then (TWAF is upper moderate).

9.2.3 Prediction of TWAF Using ANFIS for Each Type of Vane

Figures 9.5(a) to 9.5(f) illustrate the effect of initial feed concentration and membrane speed at low TMP on TWAF for different vanes, as predicted by ANFIS. The contours represent constant TWAF lines when plotted against membrane speed on the x-axis and initial feed concentration on the y-axis. The accuracy of the prediction can be evaluated by comparing it with the observations detailed by Sen et al. (171), which highlight ANFIS's performance in realistic prediction scenarios. The prediction using ANFIS concluded that vane modules with holes on the blade are efficient at low membrane speeds are also suitable for a wide range of feed concentrations. Hence, this will reduce energy consumption by maintaining low membrane speed during membrane filtration with this turbulence promoter. This observation aligns with experimental results. It may be due to the rupture of large eddies formed when fluid passes through the holes at high membrane speeds, which can carry away solutes from the membrane surface. Consequently, they fail to transport medium for local mass accumulation, ultimately reducing the TWAF. In this context, a useful conclusion is that type 1 and 2 vanes (Figures 9.5(a) and 9.5(b)) are effective at low feed concentrations with high membrane speeds. Additionally, discussing the actual hydrodynamic scenarios involving these vanes or turbulence promoters is beyond the scope of this book. However, it has been found that the fixed vanes (Figures 9.5(a) and 9.5(b)) are more effective than rotating vanes without holes (Figures 9.5(e) and 9.5(f)) in disrupting the concentration polarization layer, resulting in higher permeate flux. The inset of each figure shows the results of actual observations, which closely match the predictions made by ANFIS.

The results were also validated using ANN, a simpler pattern mapping network compared to ANFIS, to assess the prediction accuracy. However, a key advantage of using ANFIS over ANN (designed with vane type, membrane speed and feed concentration as inputs to the network) is ANFIS's ability to extrapolate TWAF beyond the experimental range after network validation. One of the intricate issues with a simple ANN is including individual vanes with their design variations in a single network and extrapolating or interpolating the results using the same network. In contrast, ANFIS incorporates such individual vane characteristics into a single network through qualitative paradigms known as "rules" and validates the prediction with high accuracy, closely resembling real observations. Figure 9.6 presents a comparison of the prediction accuracy for the validation data set using ANN and ANFIS. The figure clearly demonstrates greater accuracy with ANFIS compared to ANN, which may be attributed to the improper usage of vane types as inputs to the neural network. Additionally, a key advantage of the ANFIS network over ANN is the reduction in computation time required to design the network for individual vanes, as evidenced by the fewer iterations needed.

9.2.4 Significance of Rules in Designed ANFIS Network with Vane Geometry

This section aims to elaborate on the impact of rules on predicting TWAF based on the vane geometry using ANFIS. The critical aspect of the ANFIS lies in establishing rules that reflect the qualitative aspects of the process. Therefore, the formation of rules with precise verbose descriptions is crucial for accurately predicting experimental observations. Figure 9.7 presents an error analysis using a Box-and-Whisker diagram for the predicted TWAF from RDMM with a turbulence promoter using ANFIS. This demonstrates the impact of rules (designed based on the vanes' geometry) on predictions. In the formulation of ANFIS, the member function that converts the rules into quantitative values (from layer 1 to layer 2) indicates the probability of the rules' impact or qualitative aspects (Figure 9.3(a)) on the process outcome, following these steps:

a. Layer 2 provides the overall probability of the rules, which can also be referred to as qualitative aspects

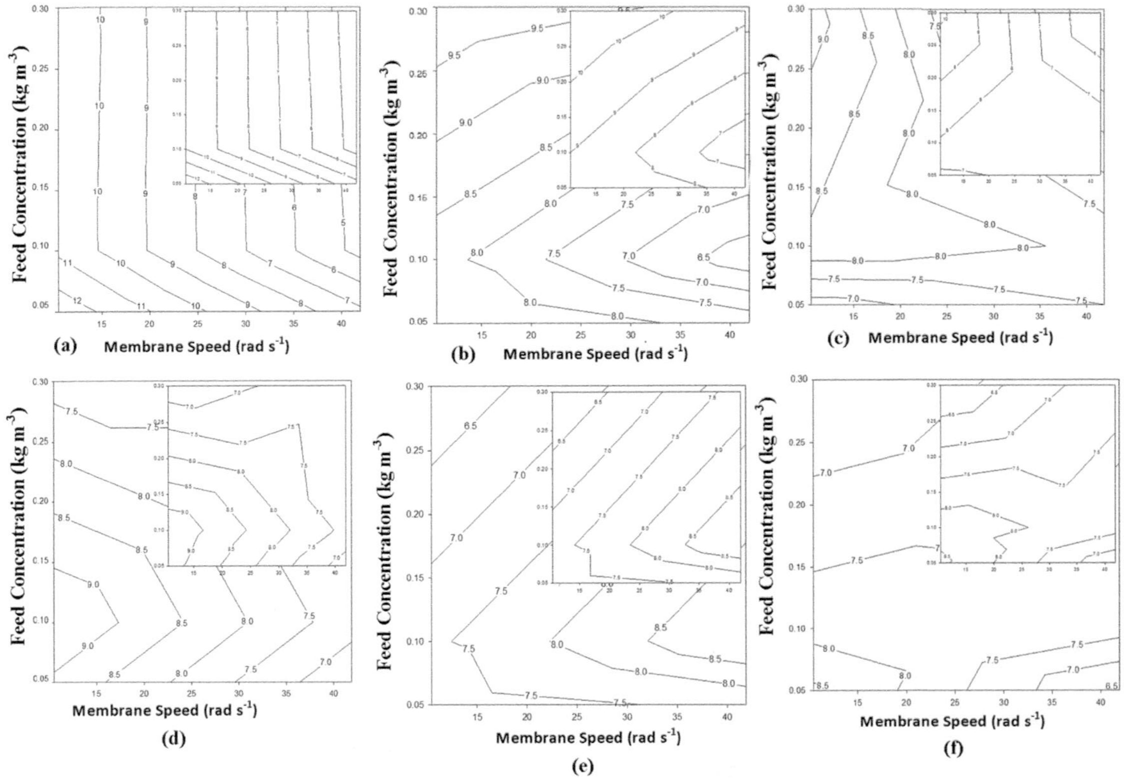

Fig. 9.5: Variation of predicted TWAF (m s^{-1}) with membrane speed (rad s^{-1}) and feed concentration (kg m^{-3}) for (a) vane type 1, (b) vane type 2, (c) vane type 3, (d) vane type 4, (e) vane type 5, (f) vane type 6 (Inset: Variation in observed TWAF (m s^{-1}) indicated on contour lines with membrane speed (rad s^{-1}) on x-axis and feed concentration (kg m^{-3}) on y-axis for (a) vane type 1, (b) vane type 2, (c) vane type 3, (d) vane type 4, (e) vane type 5 and (f) vane type 6)

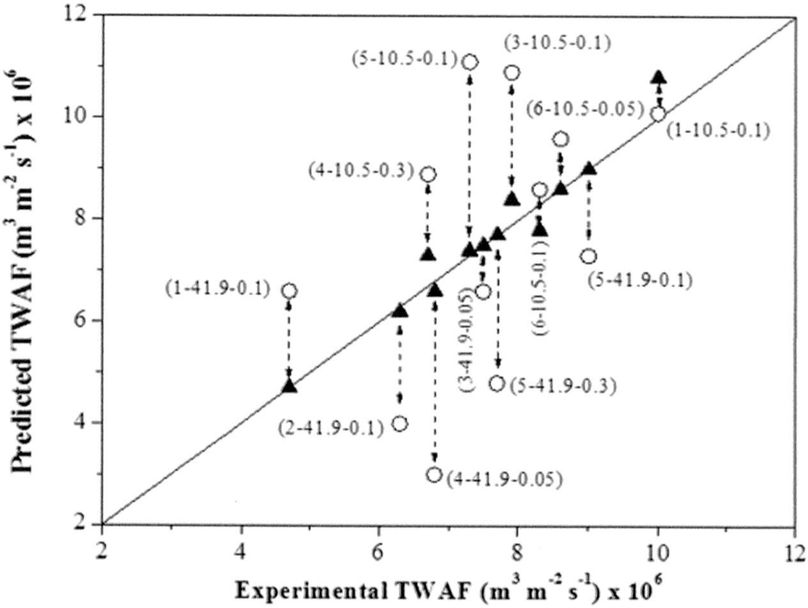

Fig. 9.6: Comparative manifestation of TWAF (m s^{-1}) predicted from validation dataset using ANFIS and ANN. Text within the parenthesis over each data point attributes to the vane type, membrane speed (rad s^{-1}) and feed concentration (kg m^{-3}) written in a respective order (▲: Predicted TWAF using ANFIS; ○ : Predicted TWAF using ANN)

b. Layer 3 indicates the significance of a particular aspect on the network outcome.

c. Layer 4 ultimately provides the prediction based on the significance of the qualitative aspect outlined in layer 3.

In the current representation using a Box-and-Whisker diagram, the extension of each quartile from the midpoint of the data set indicates the tendency of the data generated during experiments. For instance, in this case, if the upper quartile extends than the lower quartile in any plot, it suggests a higher likelihood of prediction error with the currently designed ANFIS network. The figure indicates that the rule design is considerably appropriate for turbulence promoters type 2 and 3, where predictions appear more accurate compared to other geometries. This is attributed to the extended lower quartile in the figures, which suggests fewer uncertainties during the experiments. It refers to the percentage of error that falls below the median of the percentage absolute error generated due to the applied rules in ANFIS. Despite this, the likely error is significantly higher in these cases compared to others, with errors for type 2 and type 3 vanes around 6.2% and 8.9%, respectively. However, for all the other vanes, the predominance of upper quartile over the medium an idea of increasing trend of error in the predicted values using ANFIS. Nevertheless, it is challenging to draw any conclusions about the effect

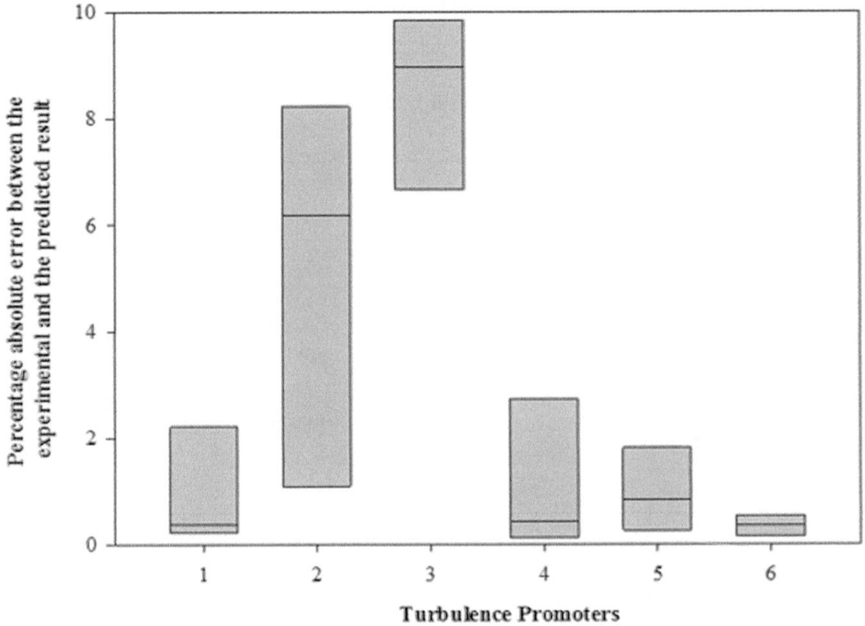

Fig. 9.7: Variation of percentage absolute error between the predicted and observed TWAF (m s^{-1}) for different vanes in the form of Box-and-Whisker diagram

of the rules solely based on the predominance of the quartiles over the median. Therefore, another parameter that sheds light on the accuracy of the network due to the rule set is the interquartile range (IQR), which is defined as the difference between the upper quartile boundary and the lower quartile boundary. The figure shows that the IQR is maximum for type 2, indicating a wide variation in error predictions due to the applied rules. For vane 6, the variation and low median demonstrate greater accuracy of the designed network. As previously mentioned, ANFIS is not merely a pattern mapping network like ANN. Instead, it attempts to capture the process concept by incorporating qualitative rules into the network alongside the pattern mapping concept. Therefore, establishing proper rules determines the probabilistic effect of each rule on the prediction (layer 3, where the weight is assigned to each independent parameter). In this case, the rules are designed based on the combination of vane types and the selected operating conditions, i.e., the range of membrane speed and feed concentration, to predict the TWAF. However, the hydrodynamic effects are not included in the model formulation, which might introduce some error in the predictions. This error margin is significant for the various rule sets (described in Table 9.3) for vane type 3 because of the holes in the blades, leading to the creation of eddy pockets and complex hydrodynamic features attributed to the vane angle. Therefore, incorporating such geometrical complexities into the predictive network as a rule will enable accurate assessment of the rule weightage (evaluated from the total probability in layer 3) on the response. Figure 9.8 presents a disparity graph that illustrates the accuracy of the prediction using ANFIS for both the training and validation data sets.

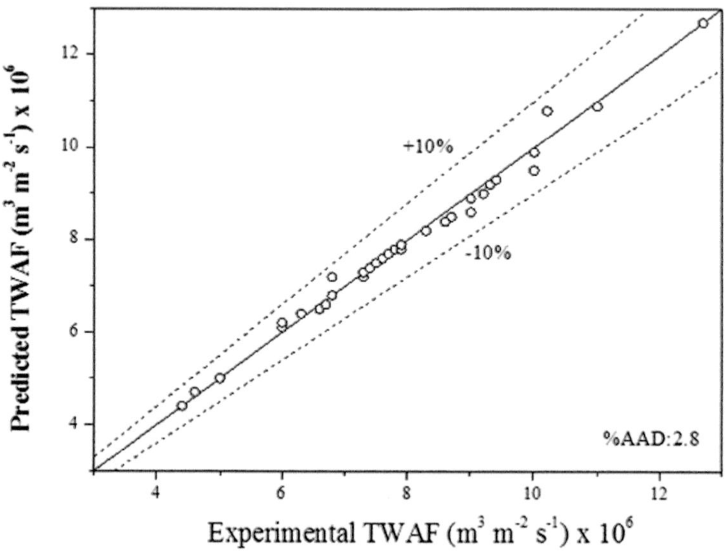

Fig. 9.8: Deviation of predicted TWAF (m s^{-1}) from the observed TWAF (m s^{-1}) for the validation and training dataset using ANFIS. The maximum and minimum deviation were considered ±10%, where the percentage AAD for the overall dataset was 2.8%

Although the current predictive model using ANFIS (Figure 9.9) lacks the ability to reduce the predictive error almost to below 5% with some turbulence promoters attached within the membrane module, the model itself is superior to ANN or any mechanistic model due to the inherent advantages of ANFIS. As previously explained, ANFIS is a pattern mapping network that combines the training-based approach of ANN with the conceptual features of the original experiments in the form of qualitative rule sets. Compared to ANN, it predicts with fewer iterations, reducing data processing time and percentage AAD. In essence, predictive accuracy of ANN solely relies on the judicious selection of training data, while in the case of ANFIS, this can be compensated by setting up proper verbose rules. However, modifying the rules with a qualitative verbose after incorporating hydrodynamics aspects based on vane geometry may lead to more realistic predictions, which will be considered as an extension to this study. Figures 9.10 and 9.11 display the response for membrane speed and concentration after compensating for disturbances using a fuzzy controller.

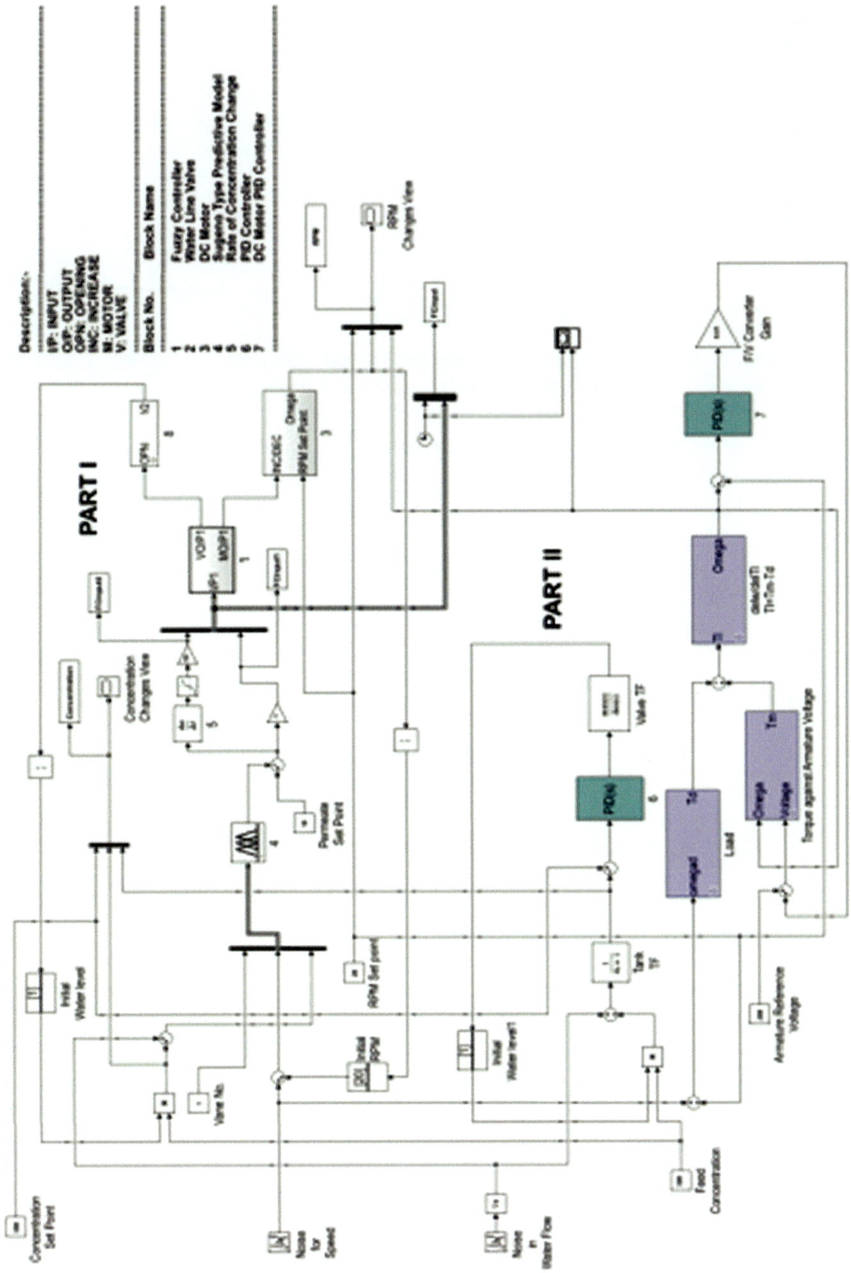

Fig. 9.9: SIMULINK model with fuzzy control

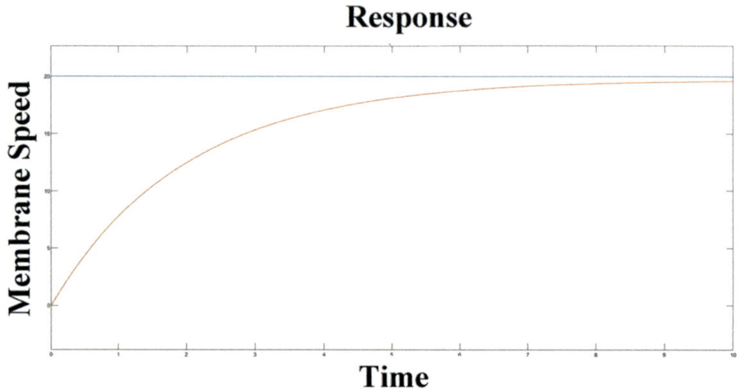

Fig. 9.10: Response view for membrane speed with fuzzy controller (Blue line: Set point; Orange line: Response)

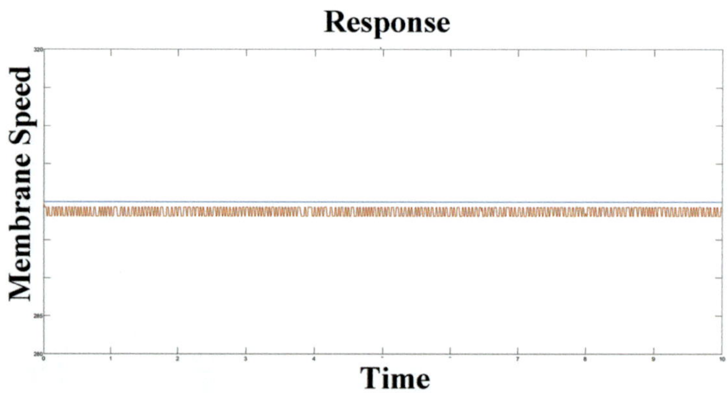

Fig. 9.11: Response view for concentration with fuzzy controller (Blue line: Set point; Orange line: Response)

9.3 Summary

- Elaboration on ANN and ANFIS architecture.
- Membrane flux predictive model development strategy with ANFIS.
- Case study on the development of predictive models for high sheared membrane modules.

Acknowledgments

We show our sincere thanks and gratitude to Prof. Chiranjib Bhattacharjee, Department of Chemical Engineering, Jadavpur University, Kolkata, West Bengal, India for his support during the collection of observations in Membrane Research Laboratory.

CHAPTER

10

Insight on Recent and Future Perspectives of Advanced Membrane Materials and Technology

10.1 Introduction: Advanced Membrane Materials and Future Membrane Technologies

In recent times, membrane-based technologies became potential candidates for a wide range of industrial and commercial applications. Their attributes, including permeability, selectivity, flux, stability, efficiency, flexibility, longevity, simplicity, compatibility with other processes, cost-effectiveness, energy intensity, easy control and scalability, position membrane-based processes as ideal solutions. The previous chapters thoroughly examined various types of membrane separation processes, their applications, and the pros and cons of each. This chapter will focus on recent advancements and the future challenges in the promising field of membrane based technologies and processes. Several novel and advanced materials, methods, and technologies are being explored in membrane science for their further development, advancement, and enhancement in terms of effectiveness and efficiency. Despite continuous developments and enhancements in this field, membrane science and technology still appears to be in its early stages. Nonetheless, advancements in specific areas such as wastewater treatment, health care, pharmaceuticals, material science, textiles, and food have shown increased potential and opportunities for a promising future in membrane science and technology. The use of membranes in healthcare, particularly in artificial organ development, is noteworthy and continues to grow. Significant efforts and inputs are also being made to develop artificial organs (such as the retina and the artificial brain) using the same membrane-based science and technologies employed in the development of artificial liver, pancreas, and kidney. Additionally, ongoing research aims to upgrade and advance traditional membrane processes, such as packaging and membrane encapsulation. Furthermore, exploring novel nanomaterials and smart materials enhanced with specified structures, properties and functions is essential for the promising development of membrane technologies. Acquiring in-depth knowledge on the mechanisms and transport phenomena is also expected to yield positive outcomes for the enhancement of these domains. The current progress along with future perspectives related to the potential domains of membrane technology is highlighted in the consequent sections of the present chapter. This will assist in the advancement and sustainable development of membrane technologies and will also facilitate the commercialization aspect.

Currently, membrane technology is contemplated as a well-recognized technology in the industrial sector on account of its gradual development and advancement. Membrane technology is nowadays utilized efficiently for multifarious applications. Nevertheless, several significant membrane applications still require extensive research and development within the membrane science sector, as current membranes are unable to perform these applications with satisfactory efficacy. As a result, the development of advanced membrane materials attributed with enhanced properties is necessary to effectively separate challenging mixtures and address modern applications in biotechnology, food technology, pharmaceuticals, and other fields. The development of membranes with significant selectivity and efficiency for separating isomers, enantiomers, or biomolecules is of paramount importance. Similarly, smart membranes and biomimetic membranes are gaining considerable attention for performing functions analogous to those of biological membranes. In this context, advanced membrane materials such as hydrogels, responsive polymers, nanotubes, and nanocomposites are extensively researched.

10.2 Responsive Polymers

Responsive polymers are rapidly substituting the conventional polymers to improve or tune the membrane properties and include a niche to the developed membranes. There are varied classes of responsive polymers depending on the stimuli in response to which the membrane's physical properties get changed. These stimuli include environmental factors like pH, temperature, ionic strength, light, etc. or external factors like chemical cues, magnetic fields, and electrical fields (174). Modulation of the membrane's physical properties in response to change in environmental factors results in the reversible variation of mass transfer and interfacial characteristics of the used membrane. Thus, researchers are showing interest in the development of smart membranes through enabling specific functions which are homologous to biological membranes. Till now, sufficient responsive polymers are not there and thus, more thrust is required for exploring novel responsive polymers. In this context, worldwide research efforts (specifically from material scientists) are focusing into this domain and continually striving for the development of novel responsive polymers. Moreover, development of nanocomposites and copolymers will add new characteristics to the existing responsive polymers.

In the previous chapter, we have discussed that membrane fouling often becomes a major limitation in case of membrane based separation processes. If we consider the case of wastewater treatment, some of the commonly utilized membrane materials are polyethersulfone (PES), polysulfone (PS), polyamide, and cellulose acetate. In spite of possessing good mechanical, chemical, and thermal stability, these membrane materials are highly hydrophobic in nature and are prone to fouling. For handling this limitation, several cleaning methods including backflushing, chemical cleaning, air sparging, permeate backpulsing, pulsing of the feed, etc. are commonly adapted. However, these cleaning methods also involve an additional operating cost of the separation system and decrease the usual lifespan of the membrane. Furthermore, the operation of the membrane based process gets interrupted for carrying out cleaning procedures. In this context, the development of fouling-resistant membrane surfaces that adhere to the feed stream could be a highly efficient method to reduce fouling, as well as decrease the severity, frequency, and duration of cleaning procedures. Recently, many research works have explored the use of responsive smart membranes to enhance fouling resistance and reduce fouling intensity. Among these studies, most focus on developing temperature, ionic strength, and pH-responsive membranes. When environmental stimuli like temperature, ionic strength, or pH are altered, the responsive groups typically grafted onto the membrane surface change their conformation. This shift between a hydrophobic and a more hydrophilic state can lead to the desorption of pre-adsorbed solutes. For instance, an intriguing study by Himstedt et al. (175, 176) developed magnetically responsive brushes and then grafted them onto the surface of commercially available NF membranes. It has been observed that, in the presence of an oscillating magnetic field, these grafted polymer chains begin to move, creating a mixing phenomenon at the membrane-fluid interface, which in turn reduces fouling propensity. Unlike conventional membrane surface modifications that primarily aim to reduce fouling by increasing membrane hydrophilicity, this study introduced a novel approach by designing a responsive layer to induce low Reynolds number mixing at the membrane-fluid interface. Therefore, responsive membranes have become a compelling option in the field of water treatment, as they can modify the permeation flux rate and address fouling issues.

10.2.1 Temperature-Responsive Polymers

With the change in solvent quality, dimensions of polymer chains become perturbed either through expansion or disintegration of polymer chains. The variation in solvent quality can be achieved directly by adding co-solvents or indirectly by altering the system's temperature. The adoption of an indirect strategy is rather common for inducing stimuli-responsive transformations within polymer-solvent systems. This results in the display of lower or upper critical solution temperatures. Generally, the LCST (Lower Critical Solution Temperature) and UCST (Upper Critical Solution Temperature) represent the lowest and highest temperatures, respectively, on a temperature-composition diagram of a system exhibiting two-phase behavior. However, in some systems, the UCST is lower than the LCST. In such systems, when the temperature is increased from below the UCST to above the UCST, the system transitions into a single homogeneous phase, and the polymer chains become swollen. The reverse behavior is observed in systems exhibiting LCST. As the temperature increases from below LCST to above LCST, the system separates into two distinct liquid phases to minimize its Gibbs energy. In this case, the polymer chains collapse. Through molecular design, it is thermodynamically possible to control the temperature at which the transition occurs between the extended (coil) and collapsed (globule) states. For example, the LCST can be increased by adding hydrophobic co-monomers, while the addition of hydrophilic co-monomers, has the opposite effect. Additionally, the LCST of temperature-responsive polymers can be modified by adjusting the polymer chain density and average molecular weight through grafting onto or from a surface. Thus, synthesis offers promising opportunities to fine-tune the temperature.

10.2.2 Polymers Responsive Towards Ionic Strength, pH, and Light

The polymers used in the fabrication of responsive membranes are not always neutral. For instance, polyelectrolytes (PELs) contain ionizable groups. The interactions of these polymers are partly evaluated by the degree of dissociation (f) of the ionizable groups. Strong PELs typically do not show much sensitivity to the pH of the solution due to their higher degree of dissociation (f). However, at relatively higher salt concentrations, when the ionic strength of the solution approaches that within the PEL, electrostatic screening can result in conformational changes. Conversely, weak PELs respond to variation in ionic strength and external pH. As a result, they may undergo unexpected conformational changes in response to these stimuli. Weakly basic PELs expand when the pH decreases, whereas weakly acidic PELs expand with an increase in pH. When it comes to variations in ionic strength, weaker PELs tend to disrupt at higher ionic strengths due to the effective screening of similar charges along the PELs.

Furthermore, the presence of photochromic units like spiropyran, azobenzene, viologen, diarylethene, etc. induces reversible photo-isomerization reactions due to light absorption. Reversible photo-isomerization results in transition between two states of photochromic components. This results in molecular changes in size, color, and group polarity, which can then be observed as macroscopic property changes after amplification. For instance, the permeabilities of membranes containing photochromic viologen groups can be reversibly controlled through redox reactions. The viologen components possess two distinct redox states. After the treatment of viologen groups with a reducing agent (sodium hydrosulfite solution), reversible reduction of these moieties to the radical-cationic state takes place from the di-cationic state. Usually, viologen components in the di-cationic state show high solubility in aqueous phase. However, a reduction in their solubility is observed when they exist in the reduced radical-cation state. Thus, when the grafted viologen is in the di-cationic state in viologen moieties grafted membranes, the polymer chain may be extended by the charges present on the side chains, expanding deeper within the pores resulting in lower permeability. On the other hand, when the viologen polymer grafted in the membrane is reduced to its radical-cationic state, the hydrophobic radical chains become more collapsed or entangled, thereby increasing permeability.

Most of the previous literature focused on the use of a single responsive mechanism. Few studies have explored membranes tuned by either mixed polymers or block copolymers, where each polymer responds to a distinct stimulus. Block copolymers and mixed polymer brushes exhibit switching or adaptive characteristics due to reversible micro-phase segregation among the distinct functionalities under varying environmental conditions. For example, individual polymers can change their surface energetic states when exposed to different solvents. By incorporating a combination of multiple independent stimuli, such membranes offer a more promising permeability response as compared to membranes grafted with a single type of polymer.

10.2.3 Responsive Mechanism

Stimuli-responsive membranes leverage the interactions between the pore structure and changes in the polarity/reactivity/conformation of responsive polymers or functional groups located either on the membrane surface or within the membrane bulk. Such transformations in uniquely tailored polymer systems are being employed in various devices and systems to support those applications that require reversibly tunable material properties. Therefore, designing novel membranes that demonstrate responsiveness in the form of physicochemical changes in response to environmental stimuli. This responsiveness typically occurs in two steps. The first step involves the utilization of stimuli for triggering the specified conformational transformations at a microscopic level. In the second step, these conformational transformations are amplified to a macroscopic level that leads to measurable variations in the membrane performance properties.

The stimuli-responsiveness of membranes can be explained based on the phase transition mechanisms of the polymers used as membrane materials within controlled environmental conditions. Phase transition mechanisms can be induced by factors such as temperature, solvent quality, type or concentration of ions, and other chemical or physical interactions. Responsiveness results from the transitions of a polymer between two states, represented by metastable energy minima for that particular system. These transitions require energy, which is supplied by the input stimulus. Every polymer is sensitive to its immediate environment, responding to external stimuli to some extent by altering its conformations along the backbone, branches, end groups, or segments. These transitions typically involve protonation–deprotonation, rearrangements of hydrogen bonding, realignment of cis–trans structure, order–disorder, or complexation–disengagement transitions. Therefore, smart membrane systems with enhanced stimuli-responsive properties can be designed by modifying the chemical composition, polymer chain length, topography, and architecture. The responsive mechanisms of polymers primarily depend on changes in polymer entropy, surface energy, and segmental interactions. Surface energy drives surface-responsive reorientation, as systems inherently tend to minimize the interfacial energy between the polymer (membrane material) surface and its immediate environment.

10.3 Aptamers (Oligonucleic Acids)

Stimulus-responsive materials change their properties when exposed to stimuli in varied manners. In the first case, these materials can undergo a bulk physical change, such as shape memory polymers do when exposed to temperature variations. In the second scenario, they can adjust their bulk physico-chemical properties, as seen in ionic liquids with switchable polarity when exposed to gas. In the third case, these materials may partially incorporate responsive segments within a non-responsive support structure. In this case, they may undergo changes in both physical and physicochemical properties, as observed with shrinkable polymer brushes exposed to light. The key difference between the third scenario and the previous two is the bulk response. Fundamentally, only a local, specific, and selective action of a stimulus is necessary to trigger an alteration in the responsive section of the material rather than the bulk material. While the responsive action is localized, it is often applied to the entire bulk material as well. Therefore, developing localized responsive actions would be beneficial, as it would make the process more energy-efficient and limit the stimulus effect on the bulk material. The solution to this issue may be aptamers, which are oligonucleic acids with characteristics of molecular recognition. In nature, this type of localized responsive action continuously occurs in the human body through responsive transporter proteins embedded within the non-responsive lipid bilayer of the cell membrane. This allows for responsive actions without the need to expose the entire body to light or temperature stimuli. Similarly to how nature efficiently operates within cell membranes, oligonucleic acids aided with molecular recognition capabilities can serve as 'gatekeepers' in porous artificial membranes. They can induce localized modifications in the release of solute particles or membrane permeability.

'Aptamers' are basically oligonucleic acids with molecular recognition abilities. These single-stranded RNA or DNA oligonucleotides are selected *in vitro* to adhere to specific targets with high selectivity and specificity. In nature, they function as a part of riboswitches, binding to small molecules in mRNA and participating in gene expression. Given the high specificity of aptamers, their extensive use in bioanalytical devices and drug delivery systems is now common. Aptamers offer several advantages over other selective biomolecules, such as proteins, antigens, and antibodies, that are commonly embedded within bioconjugated membranes. Due to their chemical stability, aptamers enable direct modifications, such as the addition of appropriate linker chemistry. Additionally, since aptamers are selected *in vitro* from a vast pool of random sequences ($>10^{10}$), the likelihood of obtaining a selective sequence is theoretically high. The process can also be automated, allowing for convenient screening of specific targets. Furthermore, since the primary composition of any aptamer is always an oligonucleotide, the basic physicochemical interactions within a support structure will remain the same regardless of the target, even if the aptamer sequence varies. Therefore, aptamers open new avenues and present significant opportunities for membrane technology in various advanced biomedical and healthcare applications. As previously mentioned, aptamers can be effectively incorporated into membranes to develop enhanced membranes with precise advanced selectivity. The use of aptamers has already moved beyond the initial stages of application in membrane technology and now requires further advancement to establish them as a true membrane-tuning material. Despite having significant potential for further improvement, aptamers also come with certain complexities. The challenges in handling aptamers are related to their inconsistency and the complex management of their functional properties. Formulating the functional properties of RNA or DNA hybrids is particularly complicated. Expertise and thorough knowledge of aptamer interactions and the forces involved are essential. However, there are very few characterization techniques available for accurately characterizing the functional properties of aptamers, which poses another significant challenge. Further research and advancements are crucial in the field of DNA or RNA based nanotechnology. Researchers are actively exploring the theories of these biomolecules to enhance their understanding and maximize their utilization in advancing membrane-based technology. Additionally, there are numerous potential materials present in nature that still need to be identified.

10.4 Nanomaterials and Nanocomposites

The nanotechnology field has advanced rapidly in recent decades. Progress in nanotechnology domain has also aided other scientific domains, contributing to the development of future novel technologies. Nanotechnology plays a crucial role in membrane science and technology. By incorporating recent advancements in nanotechnology, successful membrane modifications can be achieved for improved operation and performance. In this regard, a significant research advancement is the development of graphene and carbon nanotubes (CNTs), followed by the establishment of technology for their mass production. Typically, nanofibers, carbon-based materials (such as graphene and CNTs) or metal oxides are used in conjunction with ceramic or polymeric membrane materials to enhance membrane efficacy. These nanomaterials are incorporated into membranes through grafting, coating, and blending, resulting in membranes aided with improved functions and tailored properties. Mixed matrix membranes are created by incorporating nanomaterials into a polymer matrix. These nanomaterials enhance the properties of mixed matrix membranes, resulting in higher flux rates, improved permeability, increased mechanical strength, and antifouling tendencies. They played a significant role in elevating the membrane strength and functionality. The complexities involved in utilizing nanomaterials in membrane fabrication have been extensively explored, solved, and understood by the research community. Given their significant role in membrane technology, the future looks promising for the use of nanomaterials in developing novel membranes. In the near future, advanced materials can be developed by incorporating nanomaterials with various other materials. Consequently, these nanomaterials are expected to enhance the positive properties and functionality of the developed material, making them highly suitable for use in the membrane science domain. Therefore, the future of membrane technology is inseparable from nanomaterials. It is crucial to focus current research efforts on developing and utilizing novel nanomaterials for their applications not only in membranes but also in supplementary fields. These novel nanomaterials will help increase the overall surface area while reducing the consumption of material, resources, labor, and energy. Furthermore, incorporating these novel nanomaterials by embroidering or knitting them into membrane pores or the membrane itself could yield unprecedented results.

The combination of polymers and nanomaterials creates a material known as nanocomposites, which are stronger, highly efficient, and functional. When nanocomposites are used, the resulting material or process benefits from increased strength, versatility, and longevity. The novelty of nanocomposites lies in harnessing the benefits of both polymers and nanomaterials, as their individual properties are substantially different. The synergistic effect of the properties of polymer and nanomaterial creates an advanced material with unique characteristics, making them ideal for utilization in membranes. These characteristics typically include chemical resistance, mechanical strength, reinforcement, mechanical resistance, temperature resistance, barrier properties, selectivity, electrical properties, magnetic properties, bactericidal properties, antibacterial properties, and anti-fouling properties, among others. This synergistic effect enhances the overall effectiveness of the membrane, ensuring high performance efficiency. Consequently, nanocomposites play a dual role by providing resistance to external forces and adding functional attributes to the membranes. These improvements ultimately enhance the total permeable flux, anti-fouling tendency, and longevity of membranes. The development of thin-layer nanocomposite membranes is a significant research advancement in the field of nanotechnology domain, which can be valuable in processes such as reverse osmosis and forward osmosis. These nanocomposites effectively reduce resistance and fouling tendencies to a satisfactory extent, significantly enhancing the total permeable flux rate. For polymeric membranes, the development of nanocomposites is truly a boon. However, further advancements in knowledge acquisition and technology development are necessary. The search for novel materials must be intensified. The future prospects lie in producing materials or nanocomposites from waste products such as slag, fly ash, sludge, and others generated by various industries. Additionally, interdisciplinary knowledge and collaborations can facilitate the rapid and effective advancement of the nanocomposites' domain.

10.5 Hydrogels

As mentioned in the preceding chapter, hydrogel constitutes a three-dimensional structure formed by chemically or physically cross-linked polymers. Upon the introduction of water, these hydrogels display swelling properties, positioning them as potential life-forming materials. Also, hydrogels are inherently biocompatible, featuring low interfacial surface tension and a high storage capacity. These properties make hydrogels exceptional advanced materials for creating smart membranes, which can be utilized in the biotechnology industry and the healthcare sector. There are significant opportunities for utilizing hydrogels in the fabrication of stimuli-responsive and biomimetic membranes. Hydrogels with responsive mechanisms to chemical, mechanical, light, and pH stimuli are also attracting research attention for their potential in advancing membrane technology. Nonetheless, numerous challenges still need to be addressed for the comprehensive industrial application of hydrogels. To date, there have been developmental complexities in creating a hydrogel system that can rapidly respond to stimuli of low intensity or magnitude. Furthermore, there is opportunity to synthesize a single hydrogel that can respond to different external stimuli and its subsequent usage on the industrial level for varied applications.

10.6 Hydrophobic, Oleophobic, Oleophilic, Superhydrophilic Materials for Membrane

Improper discharge of industrial organic pollutants along with oil spill accidents has severed water pollution to an alarming state. Consumption of such contaminated water is completely prohibited as this may affect vegetation, organisms, and human being staying at near proximity to the contaminated water bodies. In the last decade, membrane separation enhanced with specialized wetting behavior has gained striking attention for addressing this problem. Hydrophobic and hydrophilic membranes are especially fabricated for carrying out the aqueous phase filtration. Moreover, the lotus leaf found in nature has ignited the innovative concept of developing superoleophobic and superhydrophobic surfaces. The development, progress, and growth in the domain of colloid and interface science and bionics in the recent past have facilitated to convert the idea into a reality. The overall characteristics of these super materials are because of the surface wettability, which is controlled by the factors including topography and chemical composition of the material's surface. Various types of super materials are there depending on the raw materials utilized for their preparation, like fabric, metal, foam, sponge, and carbon-based materials.

For instance, to separate oil-water biphasic mixture, the hydrophobic and hydrophilic membranes are utilized through selective blocking of the water and oil phase, respectively. On account of lower surface energy of the hydrophobic materials, a higher contact angle is displayed by the droplets of aqueous phase with a solid surface. Oppositely, in case of the hydrophilic membranes, their surfaces get wetted by the water droplets. By tuning the chemistry and roughness properties of the surface, a specialized wetting condition is to be maintained which will be appropriate for satisfactory separation of oil-water mixture. Surfaces enhanced with intense wetting behavior (superoleophilic/superhydrophilic) or intense non-wetting behavior (superoleophobic/superhydrophobic) are developed and consequently utilized in varied industrial sectors that deal with the separation of oil-water biphasic mixture. Introduction of a concurrent superoleophilic and superhydrophobic (mesh-based) membrane was done first in the year 2004. It is well-known that a catastrophic oil-spill incident (the Deepwater Horizon oil spill) had occurred in 2010 at the Gulf of Mexico. Oil spills are those disasters that include toxic and recalcitrant components to the sea and ocean. In this unavoidable accident, leakage of around 100 million barrels of oil had occurred from a malfunctioning valve. Unfortunately, the conventional mechanical measures can only handle <10% of the surface oil. Such poor oil recovery efficiency strengthened the research efforts on the development of superoleophilic and superhydrophobic membranes and sorbents as the efficient and selective tools for satisfactory recovering of the oil from contaminated stream. Therefore, for addressing the water pollution induced by organic pollutants and oil, materials having super wettability are engineered. These materials show promising result in terms of efficiency and effectiveness during separation of oil from water.

Superoleophilic and superhydrophobic materials can efficiently separate oil from aqueous phase by easily absorbing the oil and allowing the oil to enter the membrane pores, while rejecting the aqueous phase. Contrastingly, superoleophobic and superhydrophilic materials act in a reverse manner through complete rejection of oil molecules and allowing water for permeation. Inspiration of such type of materials is derived from fish scales, which shield fishes from oil pollution to a great extent in the ocean. As compared to the superoleophilic and superhydrophobic materials, these materials' performance is usually better in terms of low fouling tendency. This attributes to the fact that the oil molecules clog the membrane pores and surface and consequent cleaning is also very hard in the case of superoleophilic and superhydrophobic materials. Depending on these materials, smart membrane can be fabricated which will be enhanced with switchable wettability. In such materials, provisions of controlling the wettability are there with the aid of external stimuli. Researches are going on the separation of oil-water emulsions through adopting all of these above-mentioned smart materials enhanced with specialized wettability.

In spite of having huge potential for the efficient separation of oil-water mixture, specialized wettability materials are still in the initial stage of research. Thus, these materials are facing criticalities and intricacies during the industrial scale operation to separate oil from oil-water mixture. Primarily, specialized wettability materials should have good stability, durability, and proper design. Secondly, the developed materials with super wettability cannot be directly utilized in the sector of oil-water mixture separation. Thus, this sector demands more advancement to facilitate the direct use of super wettability materials at the site of oil contamination or spill itself. Third, the technology for mass scale production of these materials is underdeveloped and thereby limiting their commercial applications in industrial sectors. Another limitation is the lacking of these materials in terms of rapidness and greater output of the separation process of oil-water mixture because of the contradiction in between the sizes of oil droplets and membrane pores. It is also essential to formulate techniques for realizing the separation of a wide variety of oil-water emulsions. The research studies in this field were undertaken mostly for oil-water mixtures of lower viscosities, which is far from real mixtures. Therefore, research should focus on higher viscosity oil-water mixtures. Intricacy involved in handling highly viscous oil-water mixtures is intense membrane fouling and problems in consequent cleaning. Henceforth, further development and advancement are necessary in this field for suitable utilization of super wettability materials.

10.7 Smart Gating Membranes

The evolution of smart gating membranes is inspired from the concept of ion channels existing across cell membranes. Smart gating porous membranes contain functional gates which can dramatically control or modulate the permeation behavior through these membranes in response to external chemical or/and physical stimuli. These smart gating membranes are gaining continuously increasing attention and interests from diversified application fields. Myriad applications in diversified domains involve water treatment and purification, biological or chemical smart separations, controlled release, tissue engineering, sensors, valves, etc. Responsive materials are utilized for synthesizing smart membranes that create responsive gates within the membrane pores. As a result, under the influence of any external physicochemical stimuli, including pH, temperature, magnetic or electrical field, these responsive gates of membranes response and thereby control the transmembrane transport and permeation behavior. Presently, these smart gating membranes hold a prime place in sustainable development because of their advantageous attributes like almost negligible additives and chemicals requirements, high energy efficacy, better product quality, and cost intensiveness.

The real success and versatility of these smart gating membranes are largely dependent on their characteristics such as easy fabrication, high permeable flux, prompt and effective response, high mechanical strength, and high chemical stability. These afore-mentioned qualities will affirm the overall less cost involvement, effective performance, and easy commercial production of smart gating membranes and also ensure their satisfying practical performance for varied applications. However, in present time, smart gating membranes are yet facing complications and difficulties in scaling-up the fabrication process in terms of poor flux rate, less effective response, or low mechanical strength, which are severely limiting their practical applications.

For coveting these issues, appreciable research efforts are provided to formulate diversified strategies for inclusion of responsive domains within membrane materials of smart gating membranes. Depending on the timing of inclusion of responsive domains, the preparation methods of these smart gating membranes can be mainly categorized into two types. Introduction of the responsive domains can be done into membrane materials either prior or post membrane formation. Popular strategy is to synthesize smart gating membranes by inclusion of responsive domains within membrane materials post membrane formation, i.e., either onto or into presynthesized porous membrane substrates through some specific modification techniques namely grafting, coating, or pore-filling. Though such methods establish the strong mechanical property through proper physical and chemical forces in between the responsive domains and the membrane substrate, an intrinsic conflict arises in between the responsiveness and permeation flux. Generally, with the introduction of more responsive domains, the responsiveness will be more efficient, but the flux will be reduced. Oppositely, if less number of responsive domains is included onto or into the membrane, the membrane will exhibit lower responsive property but quite higher flux rate. Furthermore, fabricating a smart gating membrane adopting this procedure is a two-step method, which is complex and sophisticated for scaling-up. For instance, in case of grafting, the membranes are initially synthesized followed by the introduction of the responsive domains over the membrane surface. This process has its inherent limitations including demand of large number of responsive domains, less probability of proper grafting, and complications in mass-scale production. Therefore, the one-step processes such as blending have been widely explored by researchers for effective and promising outputs. In this process, the inclusion of responsive domains to the membrane materials, i.e. casting solution is carried out prior the membrane formation. Generally the membranes are prepared by adopting non-solvent induced phase separation process. The prime benefit of this one-step synthesis method is that mass-production of smart gating membranes is possible with the currently existing industrial equipments or technologies. Thus, scaling up of such one-step synthesis method is easy and feasible in case of easy availability of specific responsive domains or functional polymers. Nevertheless, nonsolvent-induced phase separation has its inherent disadvantages. This process occurs very rapidly during the formation of membrane. Usually, only a few seconds is required for the completion of solidification process during fabrication of polymer membranes. Because of such promptness of the process, it is quite problematic to control properly the even distribution of the responsive functional domains within the synthesized membranes. Thus, it is often challenging to judiciously control the even distribution and proper alignment of responsive functional domains within the synthesized membranes in such a fast process. Due to this, perfect or desired gates are not always developed in the membrane pores, which limits either the permeate flux or the responsiveness. Additionally, the incorporation of responsive domains results in pore formation within membrane, which gives rise to a conflict in between the responsiveness and mechanical property of the synthesized membrane. Hence, the utilization and industrial application of blending process for smart gating membrane synthesis are greatly restricted on account of these discussed limitations. This manifests that presently huge demand is there for establishment of a fabrication technique which will be easy to execute and can ensure both high permeation flux as well as proper responsiveness of the fabricated smart gating membranes. Other phase separations processes like vapor-induced phase separation process have been studied by the research fraternities to synthesize smart gating membranes enabled with simultaneous high flux rate, strong mechanical property, and significant responsiveness. In comparison to the non-solvent induced phase separation process, this vapor-induced phase separation process is much slower that facilitates to tune and control the morphology of the smart gating membranes more efficiently. Self-assembled responsive domains are also utilized for the fabrication of smart gating membranes by adopting this method so that the mechanical strength and smart responsiveness of the membrane do not get compromised. Consequently, these responsive domains will assemble themselves within the growing pores and membrane matrix because of the energy deviations occurring during the process of phase inversion. Smart polymers including supramolecular di/tri-block copolymers, poly(acrylic acid), poly(N-vinylcaprolactam), poly(Nisopropylacrylamide), and poly(benzyl-methacrylate) are commonly utilized for fabricating smart gating membranes.

Furthermore, the nature of gating in membrane may be positive or negative. If positive gating is considered, then the performance of membrane in terms of flux or permeability will enhance with an increment in the specified stimulus. Oppositely, in the case of negative gating, the performance of membrane will be lower with an increment in a specified stimulus. The commonly utilized stimuli for smart gating membranes are pH, temperature, ionic, light, chemical, electronic, and magnetic. These above-mentioned stimuli induce the smart gating membranes to function as or imitate sensory, chemical, and biological functions. These smart gating membranes are largely utilized for ionic transports as they allow selective transport of specified ions through them. The ionic transport is very essential function in living organisms. For instance, potassium (K^+) ions are helpful biologically whereas arsenic (As^{3+}) ions are harmful for human being. So, the smart gating membranes may be utilized for controlling the selective transport of these beneficial ions in a positive manner through allowing K^+ ions passage across them and in a negative manner by restraining the passage of As^{3+} ions. Moreover, smart gating membranes may also be utilized as sensors to detect chemicals such as glucose. As it is well-known fact that glucose plays a key role as indicator in case of some serious diseases including diabetes and hypoglycemia. In such diseases, timely detection for subsequent curing of these diseases is a crucial aspect. In the similar manner, gating membranes exhibiting prompt responsiveness towards electronic, light, and magnetic stimuli can be utilized for remote sensing and control applications.

Currently, smart gating membranes find their applications at varied fields in distinguished approach. Some of the notable examples are self-adjusting diffusional permeability during controlled release of glucose, self-adjusting hydraulic permeability of the membrane in case of ethanol production, stimuli induced responsive separation (such as size dependent separation, affinity dependent separation, etc.) and self-cleaning antifouling membrane. Even after exploring to such a great extent, few drawbacks are yet to be resolved. Thus, there remain enough opportunities to utilize the current existing technology for advanced applications and technologies through either modification of the existing technologies or development of novel technologies by taking the current technology as fundamental example. Smart gating membranes are found useful not only for conventional membrane applications but also for developing advanced and novel membrane applications in diversified domains. These membranes are utilized in food, pharmaceutical, biotechnology, and health care for controlled and specific drug release, real-time efficient and effective monitoring, and production of new health care products. However, mechanism of mass transfer occurring in smart gating membrane is not completely known till date. Also, the testing of smart gating membranes does not provide satisfactory result for long term usage in industrial applications. Lack of advanced technology and in-depth knowledge are there which need to be addressed for their successful application on commercial scale. In case of application in the biomedical industry, biocompatible advanced materials are required for synthesizing smart gating membranes. Therefore, future research efforts should concentrate primarily on acquiring in-depth knowledge of separation and mass transfer mechanisms, the development of novel advanced materials, new gating functions, synthesis methods for commercial scale production, and membranes enhanced with satisfactory response property.

10.8 Membranes in Health Care: Wound Healing and Drug Delivery

Membrane based applications are gaining dominance in the health care domain. This is considered as a blooming field for further exploration, research, and revenue generation in the membrane science and technology domain. Membranes for health care purposes are generally synthesized by using biodegradable and biocompatible materials. Health care applications mainly include artificial organs, hemodialysis, tissue engineering, and drug delivery.

Occurrence of wounds is common. However, the mechanism of wound healing is quite complicated. In case of a grievous injury, the proper healing does not take place and thus leads to abnormal tissues having decreased mechanical strength. Biomaterials like chitosan and chitin are well-known for facilitating the wound healing. Henceforth, such materials are now utilized for preparing artificial skin grafts. Two major skin cells named as fibroblasts and keratinocytes are utilized for growing artificial skin grafts by applying membrane scaffolds. These artificial skins facilitate the wound healing of patients in much rapid and effective manner, especially in case of burn patients. Materials enhanced with bactericidal or antibacterial attributes are also included by enduring the mechanical strength, porosity, and wettability for complete imitation of the original natural skin. This helps the wounded patient to heal faster with a reduced chance of spreading infection, formation of scar, and requirement of skin donors or autografts. Apligraf is an example of artificial skin. Likewise, Suprathel is a biodegradable and biocompatible artificial skin. Currently, there is a huge demand for the technology advancement to facilitate the mass-scale production, to enhance the stability, and to elevate the life-period of the artificial skins. Moreover, novel materials should also be developed for the further development in this sector. It is expected from the satisfactory progress rate of research in this sector that promising outcomes will be observed in the coming days.

Drug delivery is another well-investigated and flourishing segment of membrane science having lots of potential for advancements and developments in the future. For undertaking drug-delivery, membranes are a suitable agent because of their characteristics like permeability, precise selectivity, and less expensiveness. With the progress occurring in the domain of novel polymers mainly responsive or functional polymers, the usage of membranes in drug delivery has gained momentum. These responsive polymers have the ability to respond to external stimuli which makes their utilization beneficial during the synthesis of stimuli-responsive membranes responsible for the drug delivery purpose. Primarily these membranes has achieved success and become a hit. Thus, funding from varied sources is pouring to carry out these projects. The extensive research efforts resulted in the successful development of diversified novel materials for the synthesis of membranes to be used in drug delivery. Materials discussed previously in this chapter, such as responsive polymers, nanomaterials, nanocomposites, and hydrogels are being widely studied for synthesizing membranes. The permeability, selectivity, and specificity of the developed membranes are controlled properly for appropriate understanding of the elemental principles and mechanisms behind membrane synthesis, transport phenomenon, and fouling propensity. Presently, substantial emphasis is provided on the evolvement of biodegradable and biocompatible membranes for performing drug delivery. Furthermore, exploration of lab-on-a-chip technology is also carried out to make the drug delivery system more effective, efficient, and non-complicated.

Nevertheless, the present membrane drug delivery system is facing some hurdles including expensiveness, less accuracy, inadequate supply of materials, lower feasibility of up-scaling and mass production. Hence, advancement of the membrane based drug delivery systems is necessary for overcoming these challenges of the membrane technology. Rigorous in-depth studies and analysis are still necessary for proper understanding of the underlying transport phenomenon and mechanisms in a membrane based drug delivery system for improved performance. Detailed understanding of the interacting effects of the drug with the materials (polymers) and the corresponding drug stability are also quite necessary. The target of the membrane based drug delivery systems is to deliver optimum amount of drug to a specified target. For accomplishing this, advanced monitoring device and administration process enhanced with specific and efficient targeting feature will be required. In this context, if upgraded drugs having improved stability and lifetime can be developed, it can surely facilitate this industry to reach a next level.

10.9 Membrane Sensors

Sensors hold an important place in several applications for monitoring and regulating varied parameters. The immense demand for high quality sensors has lead to the introduction of advanced ultrasensitive sensors. The in-depth knowledge and analysis of the materials' microstructures and nanostructures has facilitated in the development of such rapid responsive sensors. Sensors are made by adopting varied chemical methods utilizing the knowledge of electronics and chemistry. Membranes can also be utilized as sensors because of the properties such as high available surface area, selectivity, less expensiveness, and improved analytes' protection.

In recent years, the focus is primarily to develop sensors possessing high surface-to-volume ratios along with decreased cross-sections. One dimensional (1-D) sensors are widely popular currently. Especially for this purpose, electrospun nanofiber mats may be a promising option on account of their large surface-to-volume ratios along with decreased cross-sections. The higher surface area helps by providing ample amount of adsorption or reaction sites for the analytes. Moreover, the decreased cross-section and short length facilitates the sensor by transporting an electron or a charge across the length of fiber more easily and rapidly. Additionally, the high interconnectivity and porosity are also some suitable characteristics that can further help in the development of an advanced and efficient sensor. Thus, ultrasensitive membrane sensors are majorly fabricated by utilizing electrospinning technique. As previously explained, the large surface area and high porosity result in the greater productiveness of electrospun membranes. The electrospun membranes are usually immobilized with the inclusion of responsive polymers, conductive polymers, enzymes, metal oxides, etc. to detect liquids, gases, temperature, light, pressure, electromagnetic components, and biological agents.

Vast scope is there in future for further developments in the domain of membrane sensors. For instance, as immobilized entities like metal oxides or enzymes, are not so stable, research emphasis is required to enhance these entities' stability so that the longevity of membrane sensors get enhanced. Moreover, in many cases, the response of prepared membrane sensors is not convincing which also requires further research, exploration and advancements. Membrane sensors should be improved for the precise and accurate detection of varied toxic compounds existing in different states and forms. In the similar way, there are many domains that demands further developments and improvements to make the use of membrane sensors widely acceptable. Thus, a bright future is there for the membrane sensors and accordingly, numerous works are continuing on this field.

10.10 Biomimetic and Bioinspired Membranes

Learning from nature is always an well-accepted practice among technical and scientific communities. In present era, it is quite established to explore, follow, and adopt from nature. Imitating a natural thing which is already working efficiently may enhance the chances of successfully recreating a homologous process or material. Thus, nature is always teaching us something novel every now and then. Numerous instances of the development of novel and unique technologies by following nature are there for successfully carrying out varied applications. Likewise, through exploration of nature, membrane researchers have imitated a few specific functions of nature and incorporated these functions into the newly developed membranes to tackle various intricate problems involved with membranes, especially fouling propensity. In this context, bioinspiration and biomimetics have evolved as a budding research forepart not only in the membrane field but also in the domain of chemistry, chemical engineering, and materials science. Basically, they act as a connecting bridge amidst basic science disciplines and applied engineering field. The first introduction of the term 'biomimetics' was done in the 1960s. It primarily deals with the biological systems in which their structures and functions are explicitly studied. Moreover, these systems are also considered as models to design various engineering solutions. Biomimetic membranes are mostly developed through imitating concepts existing in nature. However, bioinspiration membranes are not only restricted in imitating natural concepts. These membranes also utilize the recent advancements of engineering and technology field. This manifests directly that the scope of bioinspired membranes is broader as compared to the biomimetic membranes. Although, considering their characteristics and synthesis method, the progress of bioinspired membranes is lagging behind as compared to biomimetic membranes. The fundamental issue associated with bioinspiration and biomimetics is that in spite of easiness in exploring, sometimes it becomes tough to implant them because of technological impediments.

Both biomimetic and bioinspired membranes are synthetically developed most of the times by emulating the cell membranes, including lipid bilayers, surface layer proteins, biological antifouling surfaces, ionophore-based membranes, and also characteristics of some organisms, like such as mussels and gecko. Nevertheless, several shortcomings are still there for biomimetic and bioinspired membranes to be considered as next generation membranes. Henceforth, extensive research efforts and development is required for complete exploring and utilizing this domain. Inter-disciplinary technological advancements are essential to provide membrane researchers enough options for imitating the natural phenomenon in case of synthetic membranes. This demands in-depth and thorough knowledge of the existing biological systems to properly analyze their inherent working principles and control strategies. The groundwork on biomimetic will act as the foundation for the development of bioinspired technology. The technological advancement should allow the imitation to be perfect for deriving the raw benefits of the natural phenomenon in terms of membrane permeability, selectivity, antifouling, or antimicrobial tendency. Also the development methods for synthesizing biomimetic and bioinspired membranes should involve limited materials and simultaneously the overall process expense should be less. The current processes involved for the synthesis of biomimetic and bioinspired membranes are expensive, time-oriented, and material intensive. Therefore, chemical engineers, biologists, material scientists, chemists, and physicists from worldwide should research on a common target of preparing biomimetic and bioinspired membranes in a sustainable manner.

10.11 Summary

- Concept of responsive mechanism of responsive membranes.
- Smart gating membranes.
- Utilization of membranes in health care for wound healing and drug delivery.
- Elaboration on biomimetic and bioinspired membranes.

CHAPTER

11

En-routing Machine Learning for Membrane Separation Process Optimization

11.1 Introduction

As discussed in chapter 9, one of the critical issues with any membrane separation process is their outcome prediction because of too much cross-dependencies between different parameters related to membrane and feed. Therefore to control the process outcome i.e. flux a fuzzy-based controller has been mentioned and explained in Chapter 9. The other concern with any process is to identify the process's steady state and dynamic response followed by its optimization. The problem comes here is same as with the control of the process, i.e. identification of the independent relationships between the parameters along with the model formulation with a far more realistic assumptions. However, mechanistic model formulation with such considerations is a bit difficult task, while on the contrary data driven assumption will provide a more realistic approach with the same to understand the response provided the data-outcome information holds a good accuracy. Now what is machine learning (ML)? Machine learning is a tool to allow the working of a process or perhaps more specifically a mathematical model with minimal human engagement. It is a middle layer between artificial intelligence (AI) and deep learning (DL). AI is practically also a process that impels a computer to work as human, but the domain of AI is a larger one from speech recognition task to robotics to natural language processing to planning to machine learning. On the contrary, DL is a subset of machine learning, where the learning is being done with a regression analysis such as artificial neural network (ANN). In this chapter a brief introduction will be given on machine learning and its related mathematics along with algorithm, how they will be implemented for process optimization or response study along with a very brief on its application in membrane separation technology.

11.2 Mathematical Foundation for Machine Learning

The basic understanding of machine learning is the foundation of mathematics in order to solidify the data-driven model for a process. In membrane separation, the basic criterion is identification of process parameters that actually controlling the outcome of the process. For example consider a conventional UF where macromolecular protein has been separated out and one is doing the mass balance. So in conventional separation unit one has a feed (F) stream, a permeate (P) and a retentate (R) stream. Say the component that needs to be separated has weight fraction x_F in the feed stream, x_P in the permeate stream and x_R in the retentate stream. Consider the hold-up volume of the membrane module is V. Therefore, one can have two responses – steady state and dynamic response. Any perturbation in any of the process conditions will manifest a response of the process. Equations 11.1 and 11.2 are showing the steady state and dynamic response respectively.

$$\left.\begin{aligned} F &= P + R \\ Fx_F &= Px_P + Rx_R \end{aligned}\right\} \tag{11.1}$$

$$\left.\begin{aligned} \frac{dV}{dt} &= F - (P + R) \\ \frac{d(Vx_R)}{dt} &= Fx_F - (Px_P + Rx_R) \end{aligned}\right\} \tag{11.2}$$

Now this is a mechanistic model where we need to understand the degrees of freedom to get a solution of the system. A short glimpse of degrees of freedom is the number of variables that are unconstrained. If degrees of freedom is equal to zero all the variables are constrained, while on the contrary, when it is more than zero some variables are unconstrained and when it is less than zero the variables are overly constrained. Let us explain with equation 11.1, where the number of variables is six, when the number of equations is two. So degrees of freedom are given by the difference between the number of variables and number of equations. Here it becomes four suggests that one has four unconstrained variables and can be changed abruptly to understand their effect on the other variables which is called response. For example, F and x_F are known. So one can vary the permeation rate 'P' through applying trans-membrane pressure across the membrane maintaining a proper volume reduction ratio, which may be assumed as proportional to the concentration build-up in the retentate. Therefore an additional equation 11.3 can be developed and 'P' becomes decided.

$$\frac{Rx_R}{Px_P} = \text{volume contration ratio} \tag{11.3}$$

Now the constraint is imposed on the variation of the variables and a definite solution for each of the variables can be calculated. However, we have an assumptions over here that volume concentration factor is equal to component masses within the retentate and permeate, which may not be true as some solutes may get adsorbed within the secondary layer (fouling layer) over the membrane. Now let us extend it to a data driven mathematics, where the model is experiencing the variation in the response (x_P) with the variation in the F, x_F and volume concentration ratio (volume of the feed to retentate collected). One needs not to understand the actual dependencies between interactions of the solutes with the membrane surface. The training of the network is done with the known history and the prediction from the network will be done according to the set network parameters. Now the mathematical foundation comes with the different algorithms that are used for machine learning techniques. Therefore the best way to have a brief elaboration on machine learning is to understand different subdomains under it and exploring the mathematics to implement those machine learning algorithms in process. Figure 11.1 shows the classification of machine learning, while Figure 11.2 shows the classification of machine learning algorithm. In supervised learning the network gets trained according to the target set obtained from previous experience. So in a process, the previous data is available which is fitted to a supervised learning network and the outcome can be predicted. For example in a membrane separation process one is interested to predict the rejection after several cycles of run and washing with a particular feed. So the network is trained with the data set obtained against different weight fractions, flow rates, transmenbrane pressure (labeled data) with a set target on rejection seen for the mentioned information. Henceforth, if someone will change any of the conditions the prediction can be done and the rejection can be quoted. So this is supervised learning. Now consider a situation, which is much unseen (unlabeled data) and can't be estimated with some quantitative parameters. For example, the rupture of the membrane, the sudden decrease in the flow rate for either of the feed or sweep streams. Now let us go with some more classified approach, where one can say that the rupture of the membrane manifests pressure leakage or the sudden increase in the feed flow rate outbursts the restrictor valve or the increase in the sweep flow rate decreases the separation. While these issues with the membrane can't be quantified with proper data set, these can be considered as the experience people can observe with a conclusion. Like, for a certain range in the flow rate of the sweep stream the separation is good and beyond that the

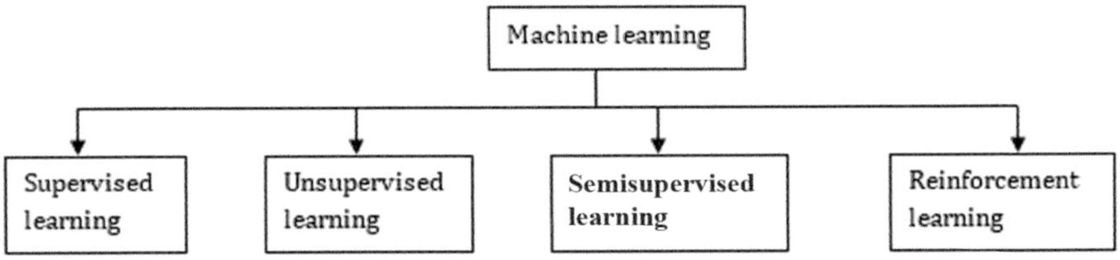

Fig. 11.1: Classification of machine learning technique (177)

separation becomes hindered. So they have a pattern to understand based on which one can actually predict the outcome from the network and accordingly actions will be considered. In this case the action may be decision on the shutting off the plant. So in one word one can say that supervised learning is the pattern mapping, while unsupervised learning is the pattern recognition.

Furthermore, there is another category called semisupervised learning where the network is minimally labeled with maximally unlabeled data (cluster identification). Semisupervised learning primarily separate the unlabeled data through proper clustering technique and labeled them to get the network trained. For example, web content identification technique, where the websites are identified and are ranked according to their hostility to banning the website. Now there are several websites which becomes a tedious task to identify individual website and to judge their hostility. Therefore, one can pick up a small group of the websites that are labeled as hostile and an unsupervised learning must be implemented to identify the pattern on the hostility based on some traits followed by labeling of the website. Now if the same concept will be extended to a process identification technique where both the supervised and unsupervised learning algorithms are required to describe the entire process. One example for that the membrane separation process, where prediction of supervised learning will not be labeled with the number of previous runs taken by the membrane, the number of washing cycle membrane undergoes, nature of the membrane, their interactions with the solutes. However, they can be clustered with some information available from the previous experience and the outcome from the membrane (like rejection or the permeate flux) can be predicted with the labeled along with unlabeled data to get a gross picture of the separation process with a particular feed. The last one is called reinforcement learning, where the learning will be on the basis of trial and error based on the accommodative scenarios developed from the environment. Planning of any process on bench scale is a kind of reinforcement learning, where lots of experiments have to be made to conclude on the basis of final possible outcome. In the following sections some of the few algorithms are elaborated that are associated with the prediction of the process outcome along with their mathematics correlating the process incoming parameters and outcome.

11.2.1 Ordinary Least Square (OLS) Technique

OLS is a supervised learning technique where a linear or nonlinear relationship is initially estimated between the incoming parameters and outcome with a regression coefficient multiplied to the individual incoming parameters. So from the experience one can actually supply the outcome against the incoming parameters.

The estimated difference between the actual outcome and the regression model outcome is estimated, the square of this difference is assigned to zero against the variation of the regression coefficient. Equation 11.4 is the estimation of the outcome with the regression model.

$$L = \frac{1}{N}\sum_{i=1}^{N} l(\hat{y}, y) = \frac{1}{N}\sum_{i=1}^{N}\left[\left(\beta_{i0} + \beta_{i1}x_1 + \beta_{i2}x_2 + \dots + \beta_{im}x_m\right) - y_i\right]^2$$
$$= \frac{1}{N}\left\|\beta X - Y\right\|^2 \quad (11.4)$$

Here, $l(\hat{y}, y)$ is the quadratic loss function calculated for N number of observations with 'm' number of incoming parameters against actual outcome 'y' and L is the average loss from altogether observations. To evaluate the regression coefficients β_{im} one has to find the minimum value of L with the equation 11.5.

$$\frac{\partial l_i}{\partial \beta_{i0}} = \frac{\partial}{\partial \beta_{i0}}\left(\beta_{i0} + \sum_{k=1}^{m}\beta_{ik}x_k - y_i\right)^2$$
$$\frac{\partial l_i}{\partial \beta_{ik}} = \frac{\partial}{\partial \beta_{ik}}\left(\beta_{i0} + \sum_{k=1}^{m}\beta_{ik}x_k - y_i\right)^2 ;k = 1, 2, \dots, m \quad (11.5)$$

||.|| is called the norm of a matrix. For minima, equation 11.5 will be assigned to zero manifesting simultaneous equations to find the values for regression coefficients. So the final matrix will take a form given by equation 11.6.

$$\left.\begin{array}{l}\begin{bmatrix}\beta_{11} & \beta_{12} & . & . & \beta_{1m}\\ \beta_{21} & \beta_{22} & . & . & \beta_{2m}\\ . & . & . & . & .\\ . & . & . & . & .\\ \beta_{N1} & \beta_{N2} & . & . & \beta_{Nm}\end{bmatrix}\begin{bmatrix}x_1\\ x_2\\ .\\ .\\ x_m\end{bmatrix} = \begin{bmatrix}y_1 - \beta_{10}\\ y_2 - \beta_{20}\\ .\\ .\\ y_N - \beta_{N0}\end{bmatrix}\\ \sum_{i=1}^{N}\beta_{ik}x_k^2 + \sum_{i=1}^{N}(\beta_{i0} - y_i)x_k + \sum_{\substack{j=1\\ j\neq k}}^{m}\beta_{kj}x_k x_j + \sum_{\substack{j=1\\ j\neq k}}^{m}\beta_{kk}x_k x_j = 0\end{array}\right\} \quad (11.6)$$

Now, let us describe a process, where separation is carried out using pervaporation for ethanol, acetone and butanol, where one needs to understand the butanol yield with the technology. The variants are considered here is the pressure, feed temperature and composition of the feed with a presumption that they are mutually exclusive in the feed. Therefore, they are fitted to a linear model and the regression coefficients are evaluated. Now this is supervised learning of the model and the model fitness is evaluated through evaluation of sum squared error (SSE) and variance (σ^2) given by equation 11.7.

$$SSE = \sum_{i=1}^{N}\left(\sum_{k=1}^{m}\beta_{ik}x_k - y_i\right)^2$$
$$\sigma^2 = \frac{SSE}{N - m} = \frac{\sum_{i=1}^{N}\left(\sum_{k=1}^{m}\beta_{ik}x_k - y_i\right)^2}{N - m} \quad (11.7)$$

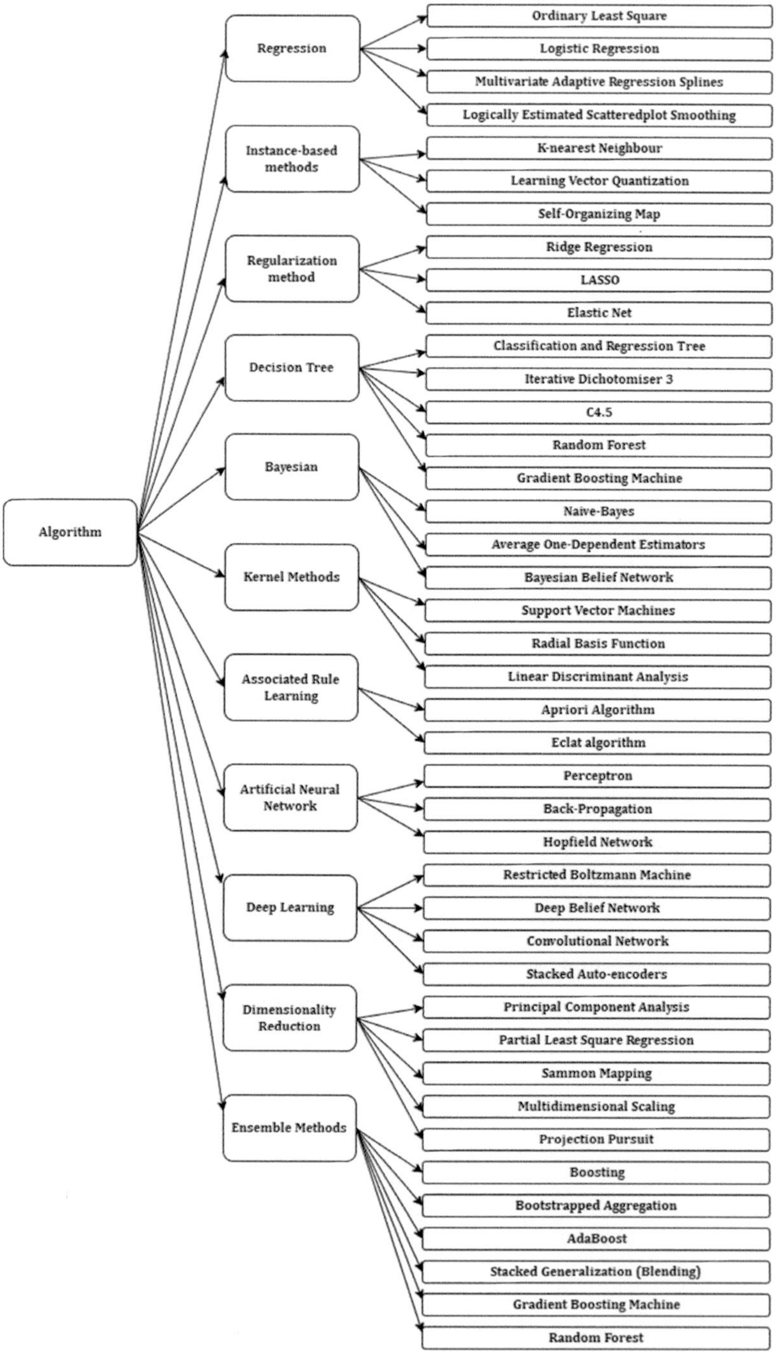

Fig. 11.2: Types of machine learning algorithms (178)

Here, β_{im}'s are already evaluated that attributes to 'm' less than the 'N' observations manifesting 'N-m' degrees of freedom.

11.2.2 Logistic Regression Technique

One of the primary achievements that are managed by OLS is the generation of continuous outcome, while in certain cases we may have classified outcome. For example, the alphabet 'A' written in different handwriting might look different, but they all are called 'A'. So different handwritten 'A' can't be treated as a continuous outcome rather it should be considered as the classified outcome in a domain. Likewise, in the example mentioned previously we need to consider even a slightest variation in the yield as a new outcome, when insignificant variation in a process outcome (not more than 5–10%) might not be considered at all. However, with the logistic regression technique such outcomes with less than 5–10% variation can be classified within a single domain and can be treated. This is true for the incoming data also.

Now before moving into the algorithms for logistic regression, let us first start with the Bernoulli's trial that is also highlighting either the success or the failure. 'p' is the probability of an event '$y = 1$' against an incident 'x', while on the contrary '$1-p$' is the probability of an event '$y = 0$' against another incident 'x'. Therefore the conditional probability for appearance of the incident y can be said as,

$$\Pr(y \mid X = x) = p^{y}(1-p)^{1-y} \Rightarrow p(y=1 \mid x) = \frac{1}{1+\exp(-y)} \tag{11.8}$$

Equation 11.8 shows the conditional probability representing Bernoulli's trials, which is also assumed by logistic regression technique attributing response in between 0 and 1. Accordingly, one needs to introduce a function that actually restricts the value in between 0 and 1 along with the classifications.

Simplifying 'p' for $y = 1$ and $y = 0$ from equations 11.8 and 11.9 respectively provides,

$$p(y=1 \mid x) = \frac{1}{1+\exp(\beta_0)\prod_{k=1}^{m}\exp(-\beta_k x_k)} \tag{11.9}$$

$$p(y=0 \mid x) = \frac{\exp(\beta_0)\prod_{k=1}^{m}\exp(-\beta_k x_k)}{1+\exp(\beta_0)\prod_{k=1}^{m}\exp(-\beta_k x_k)} \tag{11.10}$$

Now the classifier shown here is called the binomial (see Figure 11.3) where the response will be either '1' or '0'. As an example we can consider the on-off valve in any process industry. Now this could be not so straight forward as in reality one will have different possibilities in the response and it can take any value

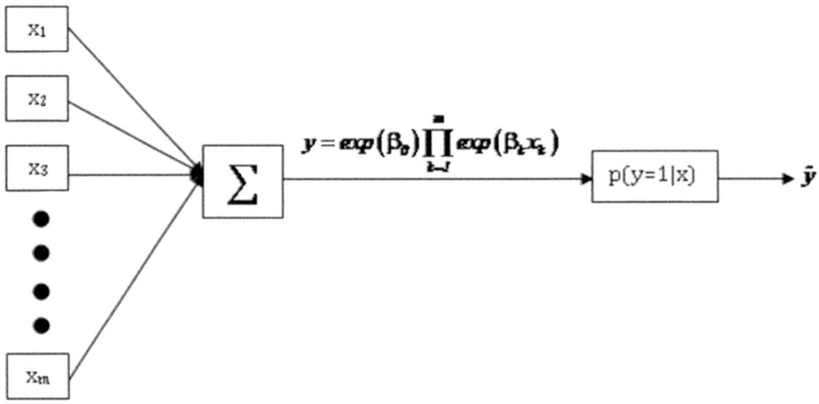

Fig. 11.3: Schematic of the supervised learning through logistic regression

within the range 0 and 1. Like in case of a flow restrictor fitted to a cross-flow membrane process, where the percentage opening of the valve directly correlates the pressure across the membrane. When one will process it using logistic regression this is called multinomial (see Figure 11.4). Now in multinomial one can identify the class '*z*' from a superclass '*Z*' for which all the outcomes will be 1 for the label '*J*'. For rest of the classes outcome will be '0'. Therefore, the probability can be written as,

$$p(y_j = 1 \mid x) = \frac{\exp(\beta_0)\prod_{k=1}^{m}\exp(\beta_k x_k)}{\sum_{j=1}^{Z}\exp(\beta_0)\prod_{k=1}^{m}\exp(\beta_{kj} x_{kj})} \tag{11.11}$$

$$p(y_j = 0 \mid x) = \frac{1}{\sum_{j=1}^{Z}\exp(\beta_0)\prod_{k=1}^{m}\exp(\beta_{kj} x_{kj})} \tag{11.12}$$

Now for a binary outcome the general form of lost function, which is also coined as the cost function once taking the average with number of classification, can be written after extending the concept from equation 11.8,

$$L = -\sum_{j=1}^{Z} y_j \log \hat{y}_j - \sum_{j=1}^{Z}(1 - y_j)\log(1 - \hat{y}_j) \tag{11.13}$$

$$\text{Cost} = \frac{1}{Z}\left[-\sum_{j=1}^{Z} y_j \log \hat{y}_j - \sum_{j=1}^{Z}(1 - y_j)\log(1 - \hat{y}_j)\right] \tag{11.14}$$

In case with the multinomial mode the outcome can be approximated with the $y_j = 1$ for a label '*J*' and thus equation 11.13 reduces to,

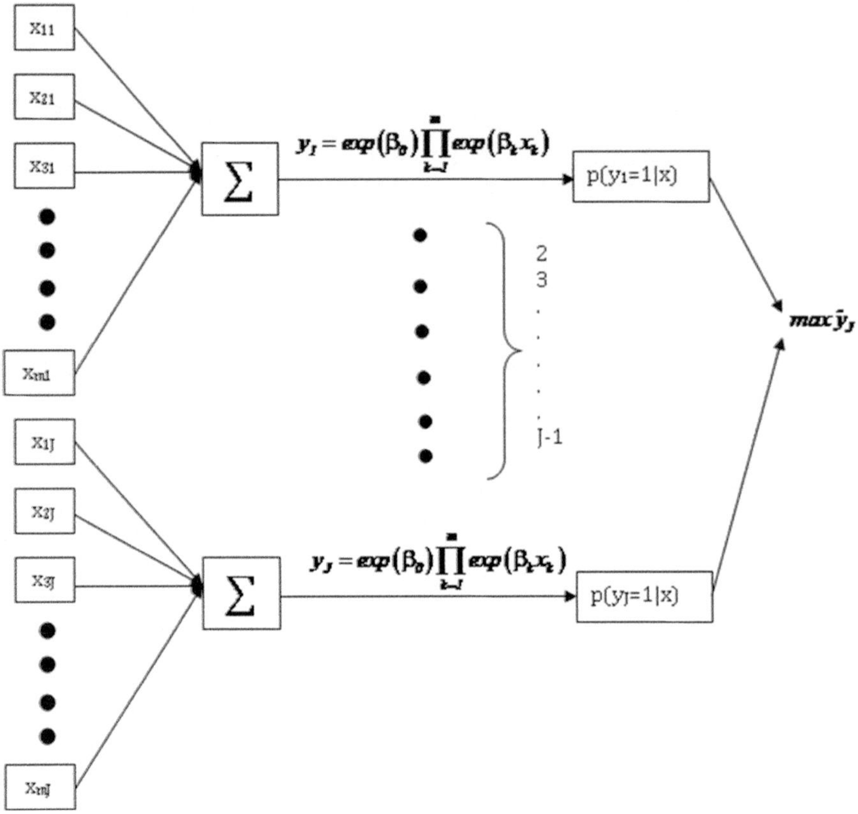

Fig. 11.4: Schematic of the supervised learning through multinomial logistic regression

$$L = -\log \hat{y}_J = -\log\left(\frac{\exp(\beta_{0J})\prod_{k=1}^{m}\exp(\beta_{kJ}x_{kJ})}{\sum_{j=1}^{Z}\exp(\beta_{0j})\prod_{k=1}^{m}\exp(\beta_{kj}x_{kj})}\right) \tag{11.15}$$

$$\text{Cost} = \frac{1}{Z}\left\{-\log\left(\frac{\exp(\beta_{0J})\prod_{k=1}^{m}\exp(\beta_{kJ}x_{kJ})}{\sum_{j=1}^{Z}\exp(\beta_{0j})\prod_{k=1}^{m}\exp(\beta_{kj}x_{kj})}\right)\right\} \tag{11.16}$$

11.2.3 *k*-Nearest Neighbor

It is the most simple and easiest nonlinear model for supervised learning with classification. The basic concept with the model is to select '*k*' nearest neighbor through measuring the distance either Euclidean (if the function is discrete) or Manhattan (if the function is continuous) followed by voting of the occurrences.

For example, say a membrane process is giving 'Good' yield, 'Moderate' yield and 'Poor' yield. So the classification is based on the yield parameter. Now for the process one has identified three operating parameters – applied pressure across the membrane (*P*), the feed concentration (*C*) and the feed flow rate (*F*). Table 11.1 shows the available data for training the network.

Table 11.1: Performance data of an UF unit

P (kgf/cm^2)	C (ppm)	F (LPM)	Yield
3	100	20	Good
2	50	20	Good
2	100	10	Poor
1	100	10	Poor
3	100	10	Moderate
3	50	10	Good

Now a new data of $P = 2.5$, $C = 60$ and $F = 11$ are supplied to predict the yield quality. The Euclidean distance calculated is given as,

$$\left.\begin{aligned} C1 &= \sqrt{(2.5-3)^2+(60-100)^2+(11-20)^2} = 41.003 \\ C2 &= \sqrt{(2.5-2)^2+(60-50)^2+(11-20)^2} = 13.463 \\ C3 &= \sqrt{(2.5-2)^2+(60-100)^2+(11-10)^2} = 40.016 \\ C4 &= \sqrt{(2.5-1)^2+(60-100)^2+(11-10)^2} = 40.041 \\ C5 &= \sqrt{(2.5-3)^2+(60-100)^2+(11-10)^2} = 40.016 \\ C6 &= \sqrt{(2.5-3)^2+(60-50)^2+(11-10)^2} = 10.062 \end{aligned}\right\} \quad (11.17)$$

Now with the selection of $k = 4$, the minimum distances are $10.062 < 13.463 < 40.016 = 40.016$ and looking at the maximum voting is 'Good'. So the yield with the new combination is good.

11.2.4 Support Vector Machine (SVM)

SVM is a linear supervised learning algorithm with classification of the data points. Now before moving into the in-depth elaboration on SVM, let us first describe a plane called "hyperplane" that actually divides the entire *N*-dimensional space into several classifications. If the space is the two dimensional one the hyperplane is the straight line dividing the domain, while in case with the three dimensional one the hyperplane is a plane dividing the domain. In case with the multidimensional space, hyperplane is the subregion dividing the domain. Now this is one the critical mathematics to solve and make a proper classification of

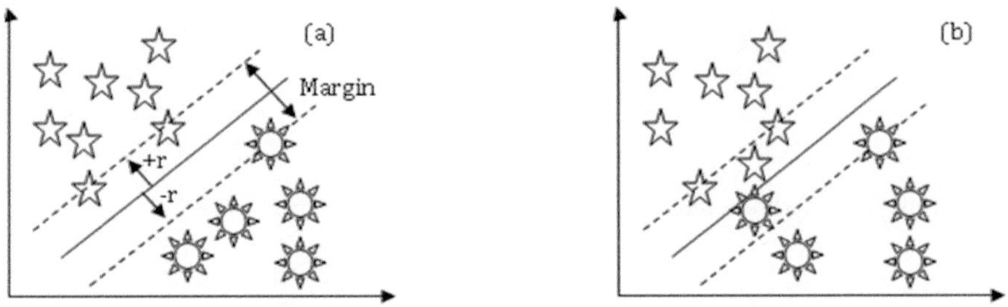

Fig. 11.5: (a) Hard margin where the support vectors are not crossing the hyperplane. (b) Soft margin where one has misclassification

the training data set. Now the classification problem can be broadly classified as the hard margin and soft margin problem (Figure 11.5).

Now in case with the hard margin linear classification the outcome is given by $Y = w^T X + B$, where 'w' is the weight and B is called the bias. During the supervised learning the weights are updated in order to train the model to predict outcome. In SVM learning for hard margin, the outcome is either +1 or –1. The problem shown here in Figure 11.5, $w^T X + B = \pm 1$. Therefore, the distance from the hyperplane (solid line) where $w^T X + B = 0$ is given by,

$$r = \frac{w^T X + B}{\|w\|} = \left.\begin{matrix} \dfrac{-1}{\|w\|} & , \forall x_i = \text{negative} \\ \dfrac{+1}{\|w\|} & , \forall x_i = \text{positive} \end{matrix}\right\} \quad (11.18)$$

$$\therefore \text{ Margin}(M) = \frac{2}{\|w\|} \quad (11.19)$$

Now it is much good realization from the Figure 11.5 that increasing the margin will ensure much prominent classification. Therefore, the optimization problem can be given by,

$$\max \frac{2}{\|w\|}, s.t.\, y_i \left(w^T x_i + B\right) \geq 1$$

or

$$\min \frac{1}{2} w^T w, s.t.\, y_i \left(w^T x_i + B\right) \geq 1 \text{ (Quadratic)} \quad (11.20)$$

Now the primal problem mentioned in equation 11.20 is dealt with the Lagrangian multiplier and the problem can be given by equation 11.21 (179).

$$L(w, B, \alpha) = \frac{1}{2} w^T w - \sum_{i=1}^{N} \alpha_i \left[y_i \left(w^T X_i + B\right) - 1 \right] \quad (11.21)$$

To find the optimal values equation 11.21 is differentiated w.r.t. w, B and ⟨ followed by making it equal to zero we get,

$$\left.\begin{aligned} w &= \sum_{i=1}^{N} \alpha_i y_i X_i \\ \sum_{i=1}^{N} \alpha_i y_i &= 0 \end{aligned}\right\} \tag{11.22}$$

Now further replacing the equation 11.22 into equation 11.21,

$$L(w,B,\alpha) = \frac{1}{2}\left(\sum_{i=1}^{N}\alpha_i y_i X_i\right)^T\left(\sum_{i=1}^{N}\alpha_i y_i X_i\right) - \sum_{i=1}^{N}\alpha_i\left[y_i\left(\left(\sum_{i=1}^{N}\alpha_i y_i X_i\right)^T X_i + B\right) - 1\right]$$

$$\Rightarrow L(w,B,\alpha) = \frac{1}{2}\sum_{i=1}^{N}\sum_{j=1}^{N}\alpha_i\alpha_j y_i y_j\left(X_i^T X_j\right) - \sum_{i=1}^{N}\sum_{j=1}^{N}\alpha_i\alpha_j y_i y_j\left(X_i^T X_j\right) + \sum_{i=1}^{m}\alpha_i$$

$$\Rightarrow L(w,B,\alpha) = \sum_{i=1}^{m}\alpha_i - \frac{1}{2}\sum_{i=1}^{N}\sum_{j=1}^{N}\alpha_i\alpha_j y_i y_j\left(X_i^T X_j\right) \tag{11.23}$$

The dual problem can be given by,

$$\begin{aligned} &\max W(\alpha) = \sum_{i=1}^{N}\alpha_i - \frac{1}{2}\sum_{i=1}^{N}\sum_{j=1}^{N}\alpha_i\alpha_j y_i y_j\left(X_i^T X_j\right) \\ &s.t. \\ &\sum_{i=1}^{N}\alpha_i y_i = 0 \\ &\alpha_i \geq 0 \end{aligned} \tag{11.24}$$

In case with the soft margin problem, a slack variable (ξ) is introduced along with a regularization parameter '*C*'. Thus the optimization statement is given as,

$$\begin{aligned} &\min \frac{1}{2} w^T w + C\sum_{i=1}^{N}\xi_i \\ &s.t. \\ &y_i\left(w^T x_i + B\right) \geq 1 - \xi_i \\ &\xi_i\left(= \max\left(0, 1 - y_i\left(w^T x_i + B\right)\right)\right) \geq 0 \end{aligned} \tag{11.25}$$

Likewise in resemblance to equation 11.24, the dual problem can be written as,

$$\begin{aligned} &\max W(\alpha) = \sum_{i=1}^{N}\alpha_i - \frac{1}{2}\sum_{i=1}^{N}\sum_{j=1}^{N}\alpha_i\alpha_j y_i y_j\left(X_i^T X_j\right) \\ &s.t. \\ &\sum_{i=1}^{m}\alpha_i y_i = 0 \\ &C \geq \alpha_i \geq 0 \end{aligned} \tag{11.26}$$

Understanding from the aforesaid discussion the existence of linear classification is a much ideal realization, while nonlinearity with the arrangement of data points is much obvious because of the mixing of the classified data sets. So the transformation from hard margin to soft margin is very obvious. There is another way through which one can convert a nonlinear classification into linear classification problem through shifting lower dimension into a higher dimensional space. This is called "Kernel trick", which is primarily mapping nonlinear classified data set to a linearly classified function (Figure 11.6).

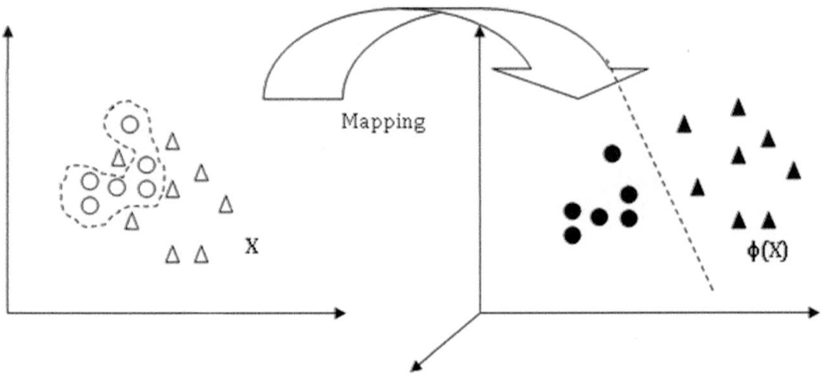

Fig. 11.6: Kernel Trick to map X into ϕ (X)

Let us first elaborate some definitions before going into the in-depth algorithm for Kernel trick. Two terms are frequently countered duality and sequential optimization problem (SOP). However both the terms are very inclusive and duality ensures SOP that is much easier to solve. In SOP the optimization is done on individual parameter sequentially and for individual optimization the previously optimized value is used. The primal problem is in this case the minimization problem w.r.t. weight (w) and bias (B). Here one needs two parameters to optimize in order to have complete set of optimized weights and biases along with the Lagrangian multipliers. This is a primal one, while the conversion of it to a single variable maximization problem is called a dual problem. In this case the single variable is $\langle_i$ that ensure the approach of equal 11.21 to a minimum one with the maximum of $\langle_i$. The Kernel trick uses the dual approach with the maximization of Lagrangian multiplier, which is given by equation 11.27.

$$\max W(\alpha) = \sum_{i=1}^{N} \alpha_i - \frac{1}{2}\sum_{i=1}^{N}\sum_{j=1}^{N} \alpha_i \alpha_j y_i y_j \phi(X_i, X_j)$$

$$s.t.$$

$$\sum_{i=1}^{N} \alpha_i y_i = 0$$

$$C \geq \alpha_i \geq 0 \tag{11.27}$$

Now the function ϕ (x_i, x_j) can take any form like linear ($\phi(x_i, x_j) = x_i^T x_j$), polynomial of degree 'd' ($\phi(x_i, x_j) = (x_i^T x_j + 1)^d$) or Gaussian ($\phi(x_i, x_j) = \exp\left(-\frac{\|x_i - x_j\|^2}{2\sigma}\right)$. The outcome in case with Kernel trick is given by $f(x) = w^T \phi(x) + B$. Now for the training of the network one needs to go for identification of the cost function given by equation 11.28.

$$\text{Cost}(E) = \frac{1}{2}\sum_{i=1}^{N}\left(y_i - w_{opt}^T \phi(X_i)\right)^2 + \frac{\lambda}{2}\|w_{opt}\|^2$$

$$\text{where, } w_{opt} = \sum_{i=1}^{N} \alpha_{i,opt} y_i X_i \tag{11.28}$$

Now to minimize the cost function given by equation 11.28 it is differentiated w.r.t. 'w' and this will eventually provides the updating rule for weight factor.

$$\frac{\partial E}{\partial w_j} = -\sum_{i=1}^{N} \phi(x_{j,i})\left(y_i - w^T \phi(X_i)\right) + \lambda w_j = 0 \tag{11.29}$$

Equation 11.29 yields,

$$\left[\phi(X_i)^T \phi(X_i) + \lambda I\right] w_j = \phi(X_i) y_i$$

$$\Rightarrow w_j = \left[\phi(X_i)^T \phi(X_i) + \lambda I\right]^{-1} \phi(X_i) y_i \tag{11.30}$$

11.2.5 Random Forest Algorithm

It is a classification problem with predicted outcome based on maximum voting. So in this particular algorithm classification is done through random sampling of the original data and the final conclusion will be made through dividing the domain or rather say partitioning the domain through making decisions from a decision tree. Now the question comes here – what is decision tree? It is a tree like models with generally three types of nodes – decision nodes (represented with square), chance nodes (represented with circle) and end nodes (represented with triangle). The decision node is where you are taking decisions, while chance nodes are saying that you may have this or that (i.e. probability of execution). End notes are execution i.e. outcome nodes. Say one is going with a membrane compaction and the person is using different membrane material for the compaction followed by the evaluation of membrane resistance. Now we all know that different membrane material manifests different maximum pressure for compaction. Assume two membranes are polyether sulfone (PES) and nylon membrane, where both are having MWCO 5 kDa. Also assume that the PES membrane can withstand 5 bar while the nylon membrane can withstand 7 bar. Figure 11.7 shows a schematic for the decision tree.

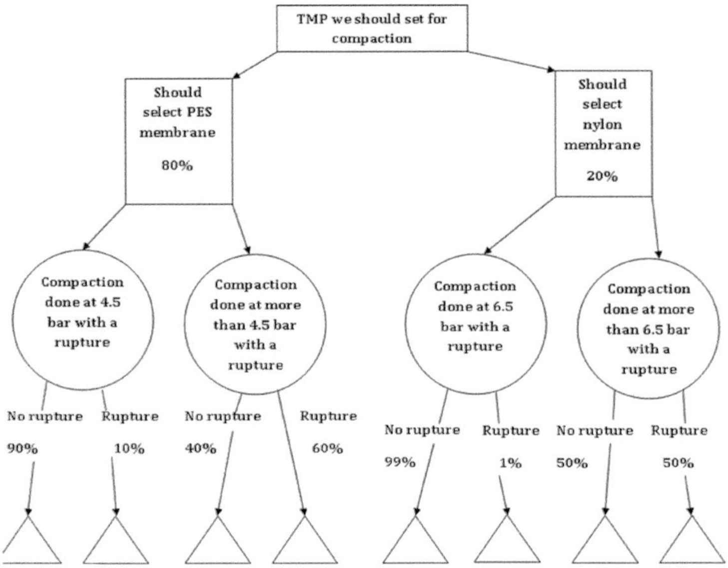

Fig. 11.7: Decision tree diagram on membrane compaction scenario

The other concept that is most vital in random forest method is the out of bag concept, where the training data set is taken through random sampling of the data after leaving some information. The idea behind the prediction is very simple, where predicting the bag-out information correctly depending on the majority voting to a specific outcome. Let us take a very well known example on fitness level. The following are the clauses that are set out.

(a) If the height is less than 180 cm and weight is less than 60 kg the person is unfit.
(b) If the height is less than 180 cm and weight is more than 60 kg the person is fit.
(c) If the height is more than 180 cm and weight is less than 80 kg the person is fit.
(d) If the height is more than 180 cm and weight is more than 80 kg the person is unfit.

It is given that the 'fit' is labeled as '1' and the 'unfit' is labeled as '0'. Figure 11.8 represents a classification process that is maintained with the decision tree diagram. Now the question is how one can represent the classification problem into a two dimensional space? First let us go with the decision tree for the same. Figure 11.9 shows the 2D representation of the classification, while Figure 11.10 shows the classification in 2D. "Decision stump" shows the barrier creating the classification.

Now as Figure 11.10 shows there are four subspaces. In S1 and S4 all the conditions are of label '1', while the other subspaces contain conditions are of

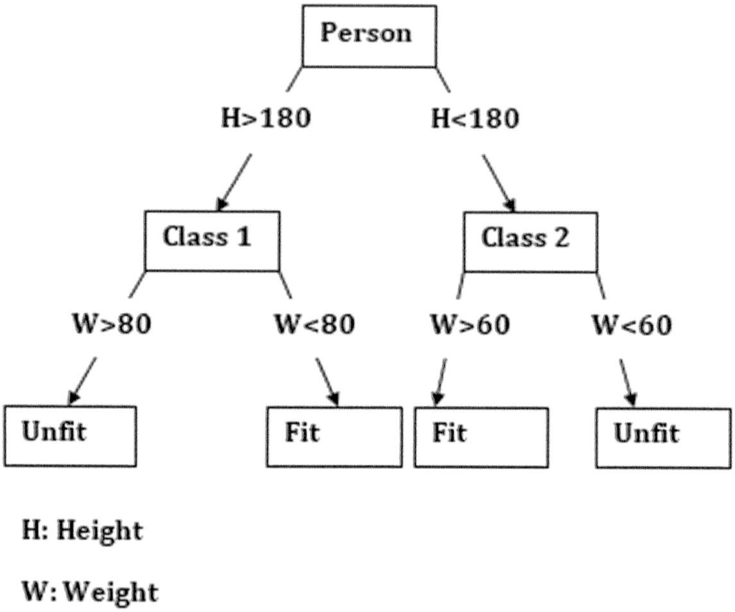

Fig. 11.8: Tree diagram with the classification for the above example

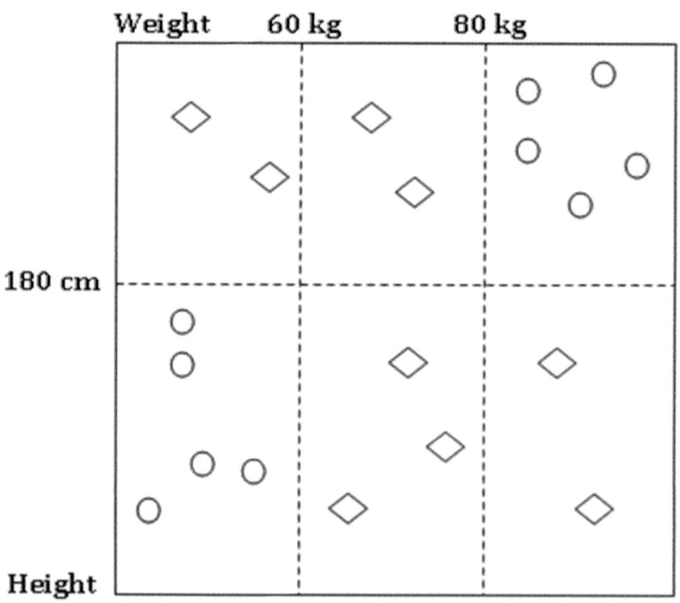

Fig. 11.9: 2D representation of the data with classification (◇: Fit(1);○: Unfit(0))

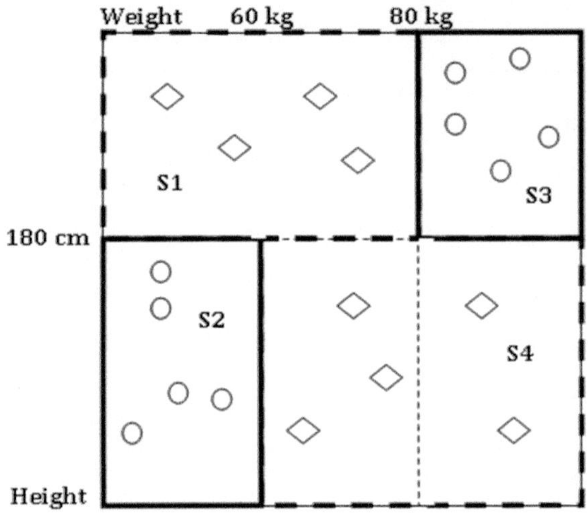

Fig. 11.10: Subspace (S1, S2, S3 and S4) creation after classification

label '0'. Now to have an effective classification, the number of labels in a domain (called "impurity") must be less and it is known as information entropy given by the following equation 11.31.

$$En(S) = -\sum_{l=1}^{L} p_l \log_2(p_l), \text{ where } p_l = \frac{N_l}{\sum_{l=1}^{L} N_l}$$

$$L: \text{Label} \tag{11.31}$$

Thus the average entropy is given as,

$$En_D(S) = \sum_{i=1}^{S} \frac{N_i}{N_{Total}} En(S_i)$$

$$S: \text{Subspace} \tag{11.32}$$

So the information gain can be approximated as $En(S)$-$En_D(S)$ implies the reduction of impurity or in other words proper classification of the information. Now this is evaluated with a particular decision tree of one sampling, which might be altered with different sampling. Out of these samplings the maximum gain will be considered. Now the split information is calculated using equation 11.33, which is further calculated to find gain ratio.

$$\text{Split}(S) = -\sum_{i=1}^{S} \frac{N_i}{N_{Total}} \log_2\left(\frac{N_i}{N_{Total}}\right) \tag{11.33}$$

$$\text{Gain Ratio} = \frac{\text{Information Gain}}{\text{Split}(S)} \tag{11.34}$$

Table 11.2: Performance data for a pervaporation unit for ethanol separation after ABE fermentation

Sl no.	Feed temperature (°C)	Membrane	Pressure across membrane (bar)	Sweep stream flow rate (LPM)	Purity of the sweep stream on ethanol (%)	Separation efficiency
1	60	A	2	20	99	Good
2	80	A	2	20	99	Good
3	90	A	2	20	99	Poor
4	60	B	2	20	99	Poor
5	80	B	2	20	99	Poor
6	90	B	2	20	99	Poor
7	60	A	5	20	99	Good
8	60	B	5	20	99	Good
9	80	A	5	20	99	Good
10	80	B	5	20	99	Poor
11	90	A	5	20	99	Good
12	90	B	5	20	99	Poor
13	60	A	2	40	99	Good
14	60	B	2	40	99	Poor
15	80	A	2	40	99	Good
16	80	B	2	40	99	Poor
17	90	A	2	40	99	Good
18	90	B	2	40	99	Good
19	60	A	5	40	99	Good
20	60	B	5	40	99	Good
21	80	A	5	40	99	Good
22	80	B	5	40	99	Good
23	90	A	5	40	99	Good
24	90	B	5	40	99	Good
25	60	A	2	20	80	Poor
26	80	A	2	20	80	Poor
27	90	A	2	20	80	Poor

Contd...

Contd...

Sl no.	Feed temperature (°C)	Membrane	Pressure across membrane (bar)	Sweep stream flow rate (LPM)	Purity of the sweep stream on ethanol (%)	Separation efficiency
28	60	B	2	20	80	Poor
29	80	B	2	20	80	Poor
30	90	B	2	20	80	Poor
31	60	A	5	20	80	Good
32	60	B	5	20	80	Poor
33	80	A	5	20	80	Poor
34	80	B	5	20	80	Poor
35	90	A	5	20	80	Poor
36	90	B	5	20	80	Poor
37	60	A	2	40	80	Good
38	60	B	2	40	80	Poor
39	80	A	2	40	80	Good
40	80	B	2	40	80	Poor
41	90	A	2	40	80	Poor
42	90	B	2	40	80	Poor
43	60	A	5	40	80	Good
44	60	B	5	40	80	Good
45	80	A	5	40	80	Good
46	80	B	5	40	80	Poor
47	90	A	5	40	80	Poor
48	90	B	5	40	80	Poor

It is much intuitive that with the classification, split gain will also increase. However, at the same time information gain will also increase that attributes to a trade-off between two gains to have a higher gain ratio. Now there could be different decision trees because of random sampling from a population. Let us discuss with the example that was discussed. In this case the sample was taken in a way that the classification is done based on 180 cm height and 60 to 80 kg of weight. Now there could be unhealthy person who are less than 180 cm height but more than 60 kg weight. This means that the classification will do some erroneous outcome that proclaims the consideration of some other parameters based on which one can go for more precision with the classification and can make the gain study. So there will be many number of decision trees those are appearing. Now for a particular decision stump arrangement one first go for maximum gain ratio (of course gain ratio must be less than 1) and finally the maximum gain ratio will be reported of all the gain ratios calculated against a particular decision stump. Now let us start with an example to get an in-depth knowledge on this. Consider an example given below with a pervaporation unit performance.

In the above problem number of "Good" (mapped with 1) is equal to 22, while number of "Poor" (mapped with 0) is equal to 26. Therefore, $p_{Good} = 22/48 = 0.46$ and $p_{Poor} = 26/48 = 0.54$. Now using equation 11.31 it can be seen the entropy is equal to 0.99. Now let us consider there are five decision stump – one called D1 for "Temperature = 80°C", D2 for "Membrane: B", D3 for "Pressure across membrane = 5 bar", D4 for "Sweep stream flow rate = 20 LPM" and D5 for "Sweep stream purity = 80%". So these are appearing as the child nodes within a decision tree. Now the child node entropy is calculated after calculation of p_{Good} and p_{Poor} for each subspace D_i (see Figure 11.11). The weighted average is calculated from equation 11.32 followed by information gain and gain ratio (equations 11.33 through 11.34). Out of those individuals, subspace gain and gain ratio maximum value for each are calculated. Now if we consider all the samples along with all the subspaces "Poor" dominates over "Good". Hence, the network output is "Poor".

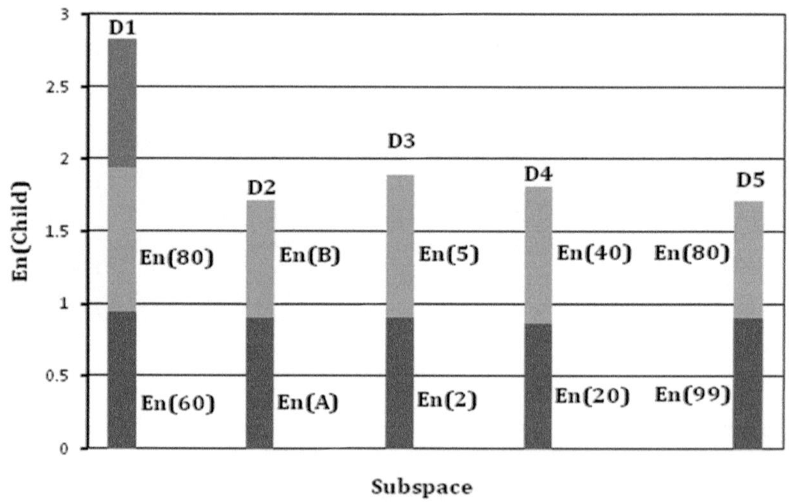

Fig. 11.11: Entropy diagram for child at different subspaces

Further, there is another indicator called Ginni index or Ginni impurity given by equation 11.35. While after the splitting Ginnia index is calculated by equation 11.36 that subsequently measure the goodness of the decision stumps (equation 11.37).

$$\text{Ginni index} = 1 - \sum_{j=1}^{M} p_j^2 \tag{11.35}$$

M is the number of classes in node space S or rather we may say in the subspace and 'p_j' is the frequency of the class 'j' in node space S.

$$\text{Ginni index (split)} = \frac{\sum_{i=1}^{S} N_i \ \text{Ginni index}(i)}{N} \tag{11.36}$$

Now, let us describe it with the example given here. Make a split on temperature at 80 i.e. one should consider "Good" separation efficiency for the temperature less than or at 80°C and the temperature more than 80°C. Now out of 32 samples for temperature less than or at 80°C, 17 samples show "Good" indicator, while 15 samples show "Poor" indicator.

$$\text{Ginni index}(\leq 80) = 1 - \left(\frac{17}{32}\right)^2 - \left(\frac{15}{32}\right)^2 = 0.9375$$

$$\text{Ginni index}(> 80) = 1 - \left(\frac{5}{16}\right)^2 - \left(\frac{11}{16}\right)^2 = 0.4296$$

Therefore,

$$\text{Ginni index}(\text{split}) = \left(\frac{32}{48} \times 0.9375\right) + \left(\frac{16}{48} \times 0.4296\right) = 0.7682$$

Now if we have started the root node as "Temperature" the Ginni index for splitting is found 0.7682. Now let us take the membrane type as the root node and check the Ginni index. With this the Ginni index found as 0.4097, which is less than the previous one indicating more pure classification and thus can be considered as the root node provided other parameters will not show the less Ginni index.

Thus trees are the basics of random forest based on which learning is happening. There is a concept called Bagging (**B**ootstrap **Agg**regat**ing**). In Bagging a random sampling has been done from the original data set and the number of decision trees will be formulated based on those random sampling deciding on the root node from Ginni index as given in equation 11.36. After the formulation of the decision tree, loss will be calculated according to either on the basis of Gain ratio or Ginni index(split). This is termed as the "Loss". Now for a particular sample of Bagging b, the loss is given by equation 11.38.

$$L_{\min}^{b} = \max_{Gain\ ratio} \sum_{i=1}^{N_b} L\left(y_i^b, \hat{f}\left(x_i^b\right)\right) \text{ or } L_{\min}^{b} = \min_{GinniIndex} \sum_{i=1}^{N_b} L\left(y_i^b, \hat{f}\left(x_i^b\right)\right) \tag{11.37}$$

There the total error in out-of-bag (OOB) is given by equation 11.39.

$$\text{Error}_{OOB} = \frac{1}{B}\sum_{b=1}^{B}\frac{1}{N_b}\sum_{i=1}^{N_b} L\left(y_i^b, \hat{f}\left(x_i^b\right)\right) \quad (11.38)$$

The outcome of the sampling will be decided based on the voting of the process. Therefore, the simplest way to write the algorithm for Random Forest algorithm can be said as,

(a) BAGGING.
(b) Formulation of decision trees with either maximum entropy gain or minimum Ginni index through splitting.
(c) Prediction of outcome after maximum entropy gain.
(d) Selection of the outcome with maximum score.

11.2.6 Naïve Bayes Algorithm

The algorithm uses Bayes' theorem to understand the possibility (or now we may call it directly probability) of occurrence of an event depending on the consideration of other feature for the occurrence. Let us say one is having a fruit bucket of apple, orange and mango. Apples are having features with color green and individually weighing around 150 g. Orange is either green or yellow and individually weighing around 100 g. Mango is all yellow and individually weighing around 200 g. So a person is buying any of the fruit with a condition is the fruit must be yellow in color. Now let us consider that the bucket has 30 apples, 20 green oranges, 20 yellow oranges and 50 yellow mangoes. Therefore the total yellow fruit in the bucket is equal to 70. Now the total number of fruit in the bucket is 120. Therefore, the probability can be defined as $P(Y) = \frac{70}{120} = 0.583$. Now say the person must pick up the yellow fruit that must be orange. So the probability of selecting yellow fruit that is orange given by $P(Y \mid O) = \frac{20}{40} = 0.5$.

Now, the Baye's theorem states that,

$$P(Y \mid O) = \frac{P(O \mid Y)P(Y)}{P(O)} = \frac{\left(20/70\right)(0.583)}{\left(40/120\right)} = 0.5 \quad (11.39)$$

Similarly one calculates the green fruit which is orange can be calculated using equation 11.39 as

$$P(G \mid O) = \frac{P(O \mid G)P(G)}{P(O)} = \frac{\left(20/50\right)\left(50/120\right)}{\left(40/120\right)} = 0.5 \quad (11.40)$$

So if some will have to choose the green or yellow orange the possibility is 50% and this is much true as they are there in equal numbers. Now if we look at the features they are mutually exclusive and will not interfere towards the selection of others. For example selection apple will not be guided by the selection of orange or mango. So the classification is much distinguished with a proper selection through application of Baye's theorem. This is called Naïve Bayes algorithm. Now if someone wants to pick up the green apple the probability will be $P(G|A) = 1$ which is higher than any other probability (and it is obvious as only category we have). This is eventually tell us that the person will surely pick up the green apple. Now in case with a classification problem, the problem is as such given as for several inputs one has a specific outcome. For example in a chemical process when two reactants are fixed, the variation in the other parameters like concentration, temperature, pressure if not widely varied, the outcome will indicate a specific product. Hence, mathematically the expression can be written as probability of outcome 'Y' given $X_1, X_2, \ldots, X_n$.

$$P(Y \mid X_1, X_2, \ldots, X_n) = \frac{P(X_1, X_2, \ldots, X_n \mid Y) P(Y)}{P(X_1, X_2, \ldots, X_n)} \alpha P(X_1, X_2, \ldots, X_n \mid Y) P(Y) \tag{11.41}$$

Furthermore,

$$P(X_1, X_2, \ldots, X_n \mid Y) = P(X_1 \mid P) P(X_2 \mid P) \ldots P(X_n \mid P) = \prod_{i=1}^{n} P(X_i \mid Y) \tag{11.42}$$

Therefore equation 11.41 can be replaced with equation 11.42.

$$\underbrace{P(Y \mid X_1, X_2, \ldots, X_n)}_{\text{Posterior}} = \underbrace{P(Y)}_{\text{Prior}} \underbrace{\prod_{i=1}^{n} P(X_i \mid Y)}_{\text{Likelihood}} \tag{11.43}$$

In a much simple way one can say that Prior is the data used for training, while posterior is the predicted outcome from the observed data. Likelihood is the Naïve Bayes classifier or in other words probability of data for given parameters. The prediction is given by equation 11.44.

$$Y_{Predicted} = \arg\max_{y} P(Y \mid X_1, X_2, \ldots, X_n) = \arg\max_{y} P(Y) \prod_{i=1}^{n} P(X_i \mid Y) \tag{11.44}$$

Let us take the example from Table 11.1, which is performance of a UF module. The outcome is yield, which is either poor or good or moderate. So basically we have three levels for the outcome. Now make a judicious arrangement of the data in Table 11.3.

Table 11.3: Arrangement of data based on yield level (example from Table 11.1)

P (kgf/cm2)	C (ppm)	F (LPM)	Yield
3	100	20	Good
2	50	20	Good
3	50	10	Good
3	100	10	Moderate
2	100	10	Poor
1	100	10	Poor

Now level for yield = "**Good**":

$P(P = 3|\text{Good}) = 0.667$; $P(P = 2|\text{Good}) = 0.333$

$P(C = 50|\text{Good}) = 0.667$; $P(C = 100|\text{Good}) = 0.333$

$P(F = 20|\text{Good}) = 0.667$; $P(F = 10|\text{Good}) = 0.333$

Now level for yield = "**Moderate**":

$P(P = 3|\text{Moderate}) = 1$

$P(C = 100|\text{Moderate}) = 1$

$P(F = 10|\text{Moderate}) = 1$

Now level for yield = "**Poor**":

$P(P = 2|\text{Poor}) = 0.5$; $P(P = 1|\text{Poor}) = 0.5$

$P(C = 100|\text{Poor}) = 1$

$P(F = 10|\text{Poor}) = 1$

$P(\text{Good}) = 0.5$; $P(\text{Moderate}) = 0.167$; $P(\text{Poor}) = 0.333$

Now let us go for a new set of parameters, where $P = 2$, $C = 100$ and $F = 20$. One wants to know the yield level. From equation 11.43 one can write,

$$P(Y|2,100,20) = P(Y)P(2|Y)P(100|Y)P(20|Y) \tag{11.45}$$

Where Y = ("Good", "Moderate", "Poor").

For Y = Good, $P(\text{Good}|2,100,20) = (0.5)(0.333)(0.333)(0.667) = 0.037$

For Y = Moderate, $P(\text{Moderate}|2,100,20) = (0.167)(0)(1)(0) = 0$

For Y = Poor, $P(\text{Poor}|2,100,20) = (0.333)(0.5)(1)(0) = 0$

Among these three levels the maximum obtained for level "Good". Hence, the outcome of these combinations will be "Good" i.e. with the new conditions the yield will be a good one. This approach is called **Maximum A Postriori (MAP)**. Now the **Maximum Likelihood Estimation (MLE)** approach tells us, $\underset{y}{\arg\max}\, P(X|Y); X = [X_1, X_2, ..., X_n]$. Therefore, according to MLE, Y = Good as it is appearing with the height probability of 0.5 (the consideration here is that for a particular level all the attributes viz. P, C and T are of equal probability). Now this is something that eventually can't be considered as we all know. So MAP is process for training that will eliminate such ambiguity appearing with MLE. Then what is the applicability of MLE? We will see that MLE is primarily use to estimate the parameters for any classifier model that are used with the Naïve Bayes. Now there are primarily three types of classifiers – Bernoulli, Multinomial and Gaussian classifier. As we can understand that Bernoulli is a binary outcome either "Yes" or "No". One may call it a binomial distribution. On the contrary, multinomial is manifesting multiple possibilities considering different combinations. A very common example is already done from Table 11.3. Now this multinomial is working when the data is discrete. If the information is continuous one is going to shift to Gaussian classifier. Now for a real process, the data must be continuous instead of discontinuous. Let us check again the previous example with the Gaussian classifier. Let us go with the "Good" classification. The average of the parameters are $P = 2.67$, $C = 66.67$ and $F = 16.67$. Further the standard deviations of the parameters are 0.577 (for P), 28.87 (for C) and 5.77 (for F). Now the probability with the test data is given by Gaussian distribution $\frac{1}{\sqrt{2\pi\sigma^2}}\exp\left(-\left(x_{Test}-\mu\right)^2 \big/ 2\sigma^2\right)$. Using the Gaussian distribution the probability on test data are given as $P(2|\text{Good}) = 1.346$, $P(100|\text{Good}) = 0.027$ and $P(20|\text{Good}) = 0.082$. Therefore,

$$P(\text{Good}|2,100,20) = P(\text{Good})\,P(2|\text{Good})\,P(100|\text{Good})P(20|\text{Good}) = 0.00149$$

One of the typical issues here with the data. For moderate we have data insufficiency and for poor we have zero standard deviations. Hence, data preprocessing is one of the primary one towards the application of Naïve Bayes. Below an algorithm is explained to apply Naïve Bayes.

1. Data preprocessing.
2. Identify the label that is considered as the outcome.
3. Training of the dataset.
4. Validation of the dataset.
5. Prediction of the outcome.

Now before concluding Naïve Bayes, we still have some issues. If we just look at the training of the dataset for the example solved using multinomial distribution it has zero probability, which is not possible in reality. With the test condition, the probability might project some value that may be too small, but can't be zero. There is technique called Laplace smoothing that will take care of zero probability. Further, one will have a term called cardinality, which is number of non-repetitive elements in a set. So let us take an example of set $A = \{1, 2, 4, 4, 5, 6\}$ and $B = \{2, 3, 4, 10\}$. Therefore, the cardinality is $7 (= \{1, 2, 3, 4, 5, 6, 10\})$. Therefore, the probability can be calculated as,

$$P(x \mid \text{condition}) = \frac{\text{Appearence of } x \text{ with condition} + \alpha}{\text{Total appearence with condition} + \alpha \times K} \quad (11.46)$$

Where, 〈 is the Laplace smoothing factor and K is called Cardinality. For the example given from the equation 11.45,

$$\left.\begin{aligned} P(2 \mid \text{Moderate}) &= \frac{0+1}{2+1\times 3} = 0.20(\alpha = 1; K = 3) \\ P(100 \mid \text{Moderate}) &= \frac{1+1}{1+1\times 2} = 0.67(\alpha = 1; K = 2) \\ P(20 \mid \text{Moderate}) &= \frac{0+1}{2+1\times 2} = 0.25(\alpha = 1; K = 2) \end{aligned}\right\} \quad (11.47)$$

Therefore, P(Moderate|2,100,20)=(0.167)(0.20)(0.67)(0.25)=0.0055.

Similarly, P(Poor|2,100,20)=(0.333)(0.333)(0.8)(0.2)=0.018

Similarly, P(Good|2,100,20)=(0.5)(0.286)(0.333)(0.5)=0.024

Here, we have highest '*P*' for the outcome 'Good'. Therefore, the prediction will be 'Good'.

11.2.7 Adaptive Boosting (AdaBoost) Algorithm

Adaptive Boosting or one may call it AdaBoost algorithm is where one may move forward towards the training of the data with '1' depth decision tree unlike random forest. Let us elaborate it in a more convenient way, where for example of enrolling PhD program based on CGPA, technical skill, English writing skill and communication skill. Table 11.4 shows the details of the example.

Table 11.4: PhD enrollment procedure

CGPA	Technical skill	English writing skill	Communication skill	PhD enrollment	Weight
>=8	Yes	Yes	No	Yes	1/8
<8	Yes	No	No	No	1/8
>=8	No	Yes	Yes	No	1/8
<8	Yes	Yes	Yes	Yes	1/8
>=8	No	No	Yes	No	1/8
<8	No	Yes	No	No	1/8
>=8	Yes	No	Yes	Yes	1/8
<8	No	No	No	No	1/8

Here one has assigned 1/8 weighting for all the information. One will go with a decision stump say for example making a decision on CGPA. If CGPA

>=8, PhD enrollment will be 'Yes', else it will be 'No'. Based on that the Table 11.5 will be,

Table 11.5: Decision stump on CGPA

CGPA	Hypothesized PhD enrollment	Actual PhD enrollment
>=8	Yes	Yes
<8	No	No
>=8	**Yes (Misclassified)**	**No**
<8	**No (Misclassified)**	**Yes**
>=8	**Yes (Misclassified)**	**No**
<8	No	No
>=8	Yes	Yes
<8	No	No

So there are 3 misclassification and 5 correct classifications. Therefore, the error generated out of the misclassification can be estimated as,

$$\varepsilon_i = \sum_{j=1}^{N} H_j w_j \tag{11.48}$$

'i' is the attribute number, 'j' is the number of observations, H_j is the hypothesis and w_j is the weight assigned. For all correct prediction $H_j = 0$ and for misclassification $H_j = 1$. On the basis of that the total error for the misclassification can be calculated as, $\sum_{CGPA} = 3/8 = 0.375$. The performance of the stump (α_i) is given by,

$$\alpha_i = \frac{1}{2}\ln\left(\frac{1-\varepsilon_i}{\varepsilon_i}\right) \tag{11.49}$$

Therefore, $\langle_{CGPA} = 0.2554$. Further, the normalizing factor required to update the weights are calculated using the following equation 11.50.

$$Z_i = w_{i,correct} \times N_{correct} \times \exp(-\alpha_i) + w_{i,incorrect} \times N_{incorrect} \times \exp(\alpha_i) \tag{11.50}$$

Therefore, updated weight can be assigned as,

$$\left.\begin{aligned} w_j(k+1) &= \frac{w_j(k) \times \exp(\alpha_i)}{Z_i} \quad (\text{for incorrect prediction}) \\ w_j(k+1) &= \frac{w_j(k) \times \exp(-\alpha_i)}{Z_i} \quad (\text{for correct prediction}) \end{aligned}\right\} \tag{11.51}$$

The update on weight will be done for each stages of attribute consideration and the next update on a particular attribute will be evaluated on the updated values of the weight for the previously considered attribute. Once all the attributes will be updated, the final weighted average for each data will be done based on the performance of the decision stump found. Equation 11.52 shows the average value calculation and based on the average value the outcome will be predicted for each data. The maximum score will be given for the test data.

$$\left.\begin{array}{l}\bar{H}_{j,\text{Predicted}} = \sum_{i=1}^{A} H_i \alpha_i \\ \text{If} \\ \bar{H}_{j,\text{Predicted}} > 0, H_{j,\text{Predicted}} = \text{'Yes'} \\ \bar{H}_{j,\text{Predicted}} \leq 0, H_{j,\text{Predicted}} = \text{'No'}\end{array}\right\} \tag{11.52}$$

Let us solve the problem given here. First one will consider the attribute "CGPA" and Table 11.5 is formulated with the hypothesis. The final updated weight after applying equations 11.48 to 11.51 results in Tables 11.6 to 11.9.

Table 11.6: Updated weight considering attribute CGPA

CGPA	**Hypothesized PhD enrollment**	**Actual PhD enrollment**	**∑**	**⟨**	**Updated weight**
>=8	Yes	Yes	0.375	0.255413	0.1
<8	No	No			0.1
>=8	**Yes (Misclassified)**	**No**			0.166667
<8	**No (Misclassified)**	**Yes**			0.166667
>=8	**Yes (Misclassified)**	**No**			0.166667
<8	No	No			0.1
>=8	Yes	Yes			0.1
<8	No	No			0.1

Table 11.7: Updated weight considering attribute technical skill

Technical skill	**Hypothesized PhD enrollment**	**Actual PhD enrollment**	**∑**	**⟨**	**Updated weight**
Yes	Yes	Yes	0.1	1.098612	0.055556
Yes	**Yes (Misclassified)**	**No**			0.5
No	No	No			0.092593
Yes	Yes	Yes			0.092593
No	No	No			0.092593
No	No	No			0.055556
Yes	Yes	Yes			0.055556
No	No	No			0.055556

Table 11.8: Updated weight considering attribute English writing skill

English writing skill	Hypothesized PhD enrollment	Actual PhD enrollment	∑	⟨	Updated weight
Yes	Yes	Yes	0.2037	0.6817	0.034884
No	No	No			0.313953
Yes	**Yes (Misclassified)**	**No**			0.227273
Yes	Yes	Yes			0.05814
No	No	No			0.05814
Yes	**Yes (Misclassified)**	**No**			0.136364
No	**No (Misclassified)**	**Yes**			0.136364
No	No	No			0.034884

Table 11.9: Updated weight considering attribute communication skill

Communication skill	Hypothesized PhD enrollment	Actual PhD enrollment	∑	⟨	Updated weight
No	**No (Misclassified)**	**Yes**	0.3203	0.3762	0.054455
No	No	No			0.230949
Yes	**Yes (Misclassified)**	**No**			0.354785
Yes	Yes	Yes			0.042768
Yes	**Yes (Misclassified)**	**No**			0.090759
No	No	No			0.100311
Yes	Yes	Yes			0.100311
No	No	No			0.025661

Now incorporating all the possibilities from Tables 11.6 through 11.9 in Table 11.10 along with the hypothesized condition an average weighted condition will be evaluated to conclude for a new prediction. The weighted average is calculated as given in equation 11.52.

Table 11.10: Decision making table

S.No.	⟨=0.255413	⟨=1.098612	⟨=0.6817	⟨=0.3762	Weighted average	Final prediction
1	Yes	Yes	Yes	No	2.035678	Yes
2	No	Yes	No	No	1.098612	Yes
3	Yes	No	Yes	Yes	1.057858	Yes
4	No	Yes	Yes	Yes	2.156471	Yes
5	Yes	No	No	Yes	0.376206	Yes
6	No	No	Yes	No	0.681652	Yes
7	Yes	Yes	No	Yes	1.730231	Yes
8	No	No	No	No	0	No

So far the descriptions or the algorithms given here is primarily the pattern classification model, where the classification is based on some of the traits associated with the process. If the elaboration will be extended to apply for any membrane separation process, all the above algorithms will deal with the basic pattern of the outputs that one is deriving from any membrane separation process. For example, if there will be a sudden fluctuation in the TMP because of some control issues but not more than ±5% variations, the flux may not vary too much and this will be considered under the same classification with the actual one. However, to predict the real values of some quantitative parameter, for example flux, one needs to apply a definite value predictive network like ANN or ANFIS as described in Chapter 9. So further elaboration will be made on these aspects, where instead of classification neural network based approach will be considered to predict the quantifiable parameters.

11.2.8 Hopfield Network

Hopfield neural network (HNN) is a fully connected network proposed by J. Hopfield in 1982. Now with the conventional neural network the primary difference here in case of HNN is its recurrent connections. So far with the conventional system, there is no feedback control mechanism, while with the HNN there is a feedback mechanism. Figure 11.12 shows HNN and it primarily follows the associative memory. Now what is associative memory? It is a technique to correlate input and output pattern with recalling the relationship later on as required. As shown, Figure 11.12 is a recurrent feedback network called HNN. Equation 11.53 shows the basic updating equation for HNN.

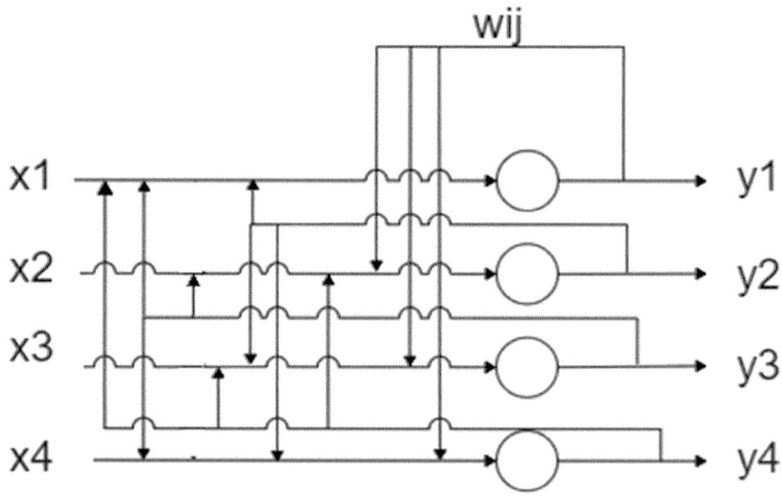

Fig. 11.12: Recurrent HNN

$$x_j(t+1) = \text{sign}\left(\sum_{i=1}^{N} w_{ij} x_j(t)\right) \tag{11.53}$$

The convergence function is given by equation 11.54.

$$E = -\frac{1}{2}\sum_{i=1}^{N}\sum_{j=1}^{N} w_{ij} x_i x_j \tag{11.54}$$

11.3 Case Studies on the Applications of ML Tools for Membrane Separation

In this section, one can find different problems approached by several researchers and the conclusion drawn by them after applying ML algorithm to solve the problem. Hence, individual subsection elaborates the problem statement along with a conclusive outcome after applying appropriate ML on them.

11.3.1 Predicting Permeate Flux for the Protein Separation with UF Membrane

Bhattacharjee et al. (180) had done a study for the separation of protein using 30kDa PES membrane with a high sheared membrane module with membrane rotation facility. pH, TMP and membrane rotation were varied to understand the separation along with an understanding of the fouling build-up over the membrane. With the fouling of the membrane the permeate flux reduces. However, the reduction of the permeate flux is compensated because of the rotation of the membrane restricting boundary layer development over the membrane surface. Each experiment was carried out for 30 mins with 20 mins post-run membrane wash in order to regain the hydraulic flux of the membrane. This eventually helps to understand the extent of the irreversible fouling of the membrane. Now in this case the development of a deterministic mathematical model that correlates permeate flux with the pH, TMP and membrane rotation is a tedious job. Also, the model may not be accurate one because of a lot idealization to obtain a solution to the model. Therefore, the authors rely on the ANN based tool to accurately predict the permeate flux. They made a comparative study with different ANN network and conclude on the fitness of an appropriate network towards the prediction. They found the maximum accuracy is with the radial basis network and least accuracy is with the feedforward network. However, with the Hopfield network they found that the accuracy is high enough and almost comparable with the radial basis network at almost neutral pH. While with the acidic pH the accuracy is less with the Hopfield. Now, why this is being happened? To elaborate the feature let us go through a more detailing on their feed system and the pH effect on the feed. They had used casein way as their feed system and the isoelectric point is 4.6. Isoelectric point is the pH at which the protein is charge neutral. Now if the environment pH is less than isoelectric pH, the protein becomes positively charged. On the contrary if the pH is more protein gets negatively charged. The membrane they had used was PES membrane which is having slight negative surface charge due to presence of $-SO_3H$ group. Therefore the interaction is more pronounced at the lower pH compared to higher pH. Now input to the network is pH, rotational speed and TMP. Rotational speed attributes to the resistance against the interaction with membrane, while TMP attributes to the force to interact. Hence these are two opposing factors and may be compensatory ones on two different mathematical frameworks. However, the other input pH is interactive, which is true with the lower pH. Now Hopfield is trying to correlate the input-output pattern, which might be difficult with the two similar counteracting conditions speed and pH on the same mathematical framework. Now this is not true with the higher pH, as the relation is co-acting. Therefore, more accuracy is with the prediction. On the contrary, radial basis function goes with the cluster configuration from the given information rather building up individual pattern correlation. Therefore, those factors are not impacting to the result and providing good accuracy in the prediction.

Odabaşı et al. (181) had elaborated a process, where the local municipal wastewater is recovered through different media and membrane filtration technology. The wastewater plant has around 12 multimedia filters, 18 activated charcoal filters, 10 UF membrane trains and 5 RO membrane trains. According to them they had predicted the salt passage, permeate flow rate and the pressure difference between feed and retentate. These informations are classified according to the feed operational parameters and feed characteristics. Feed pressure, flow rate and temperature are the feed operational parameters. Conductivity, oxidation-reduction potential (ORP), turbidity, COD and TSS are the feed characteristics. Salt passage percentage is estimated with the following equation.

$$\text{Salt passage } (\%) = \frac{\text{Permeate conductivity}}{\text{Feed conductivity}} \times 100 \tag{11.55}$$

They have compared different machine learning tool for the prediction of the parameters with prior understanding of the correlation among the input/output variables. There is a strong correlation (0.96) can be seen between feed and permeate flow rate, which might be because of mass balance. Rests of the correlations' values suggest that there wouldn't be much direct relationship between the parameters, although presence of salt manifests the conductivity of the feed and permeate. The correlation coefficient is found around 0.64. For the salt passage permeability they found that the "random forest" works better with train and test root mean square error (RMSE) of 0.71 and 0.65 respectively. As discussed a priori the accuracy in case of random forest algorithm relies more on the classification efficiency, where the Ginni values shows the purity of the classification. Figure 11.13 shows the Ginni values for different feed characteristics.

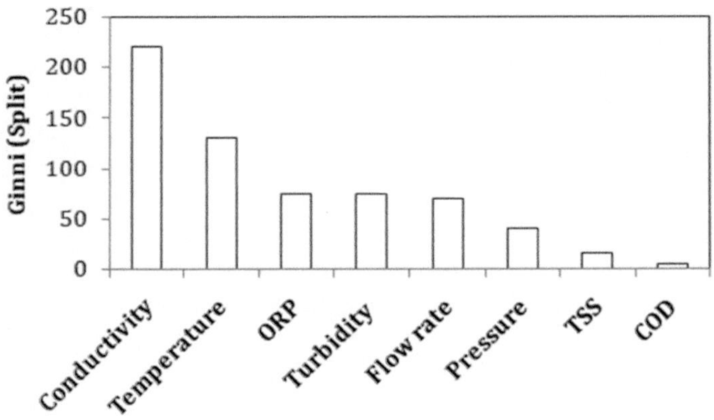

Fig. 11.13: Ginni values for different feed characteristics

From the figure, it is evident that with the conductivity the classification is good with maximum purity, while the next level of purity can be observed with the temperature. According to the authors, the temperature increase will increase the pore diameter for the polymeric membrane that eventually increases the passage of the salt. While the higher ORP values attribute to the strong oxidant potential that deforms the polyamide membrane texture followed by an increase the salt passage. Now in a brief it is understood that the direct correlation between the base classification traits and the outcome will provide better classification with higher Ginni values. Why this is happening with the random forest? Now the basis for random forest is to understand the crisp decision stump based on which the subspaces can be understood (Figures 11.9 and 11.10).

Waqas et al. (182) had predicted the membrane permeability within a membrane rotating biological contactor for the treatment of wastewater. They have correlated the effect of disk rotating speed, hydraulic residence time (HRT) and sludge residence time (SRT) with the membrane permeability using SVM. One of the typical issues with the bioreactor system is the development of proper correlation between the permeability and the operating conditions specified. They had applied both ANN and SVM to predict the permeability. Both the network shows regression coefficient value more than 99%.

Recently one of the most significant achievements that are primarily acknowledged is the amalgamation of machine language with the microfluidics calling the domain “intelligent microfluidics”. It is already being discussed that electrochemical biosensor is having a bioactive embedded membrane that detects the analytes through a biochemical reaction. So the flow of analytes within the thin channels of the device and also the diffusion through pores relates the field of microfluidics and membrane. For reference one can see a very recent review article by Antonelli et al. (183) can be consulted. Now let us take one case on glucose biosensor, where the sensor amperometric responses are predicted through different conditions like temperature, pH, benzoquinone and glucose concentration (184). They had tried to unveil the interactivity between the conditions and the amperometric response. After unveiling the interactions they have optimized the condition to maximize the signal response through ANN and SVM. One of the prime observations that were envisaged primarily is that the SVM with linear kernel function outperforms the performance, while SVM with radial basis kernel function shows more accuracy. SVM with radial basis function shows around 99% regression coefficient while linear one shows 52%. Even SVM with radial basis function is marginally more accurate than the ANN. Optimization of the conditions was done through genetic algorithm and simulated annealing, which are further used to predict the response with ANN and SVM with radial basis function.

11.4 Summary

- Brief on the types of machine learning.
- Elaboration on different supervised and unsupervised machine learning algorithms.
- Discussion on different pattern mapping algorithm.
- Discussion on autoassociative memory based algorithm.
- Discussion on different case studies with machine learning algorithm applied for membrane separation process.

APPENDIX

A

Levenberg-Marquardt Algorithm (LMA)

LMA is an optimization problem that can actually solve non-linear problem and also can be understood as the non-linear least square method leading to a minimization of the residuals. During optimization of any nonlinear problem the finding of optimum point is done according to the Newton's method given in equation A.1.

$$x_{i+1} = x_i - \left(\nabla^2 f(x_i)\right)^{-1} \nabla f(x_i) \tag{A.1}$$

Levenberg used a modification in this updating rule and given by $x_{i+1} = x_i - (H + \lambda I)^{-1} \nabla f(x_i)$, where H is the Hessian matrix given by $H = J^T(x_i) J(x_i)$. J is called Jacobian and given by $J(x) \quad \frac{\partial f(x)}{\partial}$. The basic algorithm for LM is given in the following Table A.1.

Table A.1: LMA

Steps	Statement		
1	Initial guess for 'x_{old}'		
2	$x_{new} = x_{old} - (H + \lambda I)^{-1} \nabla f(x_{old})$		
3	Check for error $\lvert x_{new} - x_{old} \rvert \leq$ tolerance		
	Yes:	Terminate the program and report x_{new}	
	No:	Check to error increase or decrease for the previous iteration	
		Error decrease:	Decrease λ by a factor 10
		Error increase:	Increase λ by a factor 10
		Go to step 2	

References

1. Loeb S, Sourirajan S. Sea water demineralization by means of an osmotic membrane. Saline Water Conversion—II. Advances in Chemistry. American Chemical Society, 1963, 38: 117-132.
2. Baker RW. Reverse Osmosis. Membrane Technology and Applications. 2nd ed. 2004, 191-235.
3. Cadotte JE. InventorInterfacially synthesized reverse osmosis membrane patent US4277344A. 1979.
4. Balster J. Plate and frame membrane module. pp. 1-3. *In:* Drioli E, Giorno L (eds). Encyclopedia of Membranes. Berlin, Heidelberg: Springer Berlin Heidelberg. 2015.
5. Balster J. Spiral wound membrane module. pp. 1-3. *In:* Drioli E, Giorno L (eds). Encyclopedia of Membranes. Berlin, Heidelberg: Springer Berlin Heidelberg. 2015.
6. Kleinheins J, Shardt N, El Haber M, Ferronato C, Nozière B, Peter T, et al. Surface tension models for binary aqueous solutions: A review and intercomparison. Physical Chemistry Chemical Physics, 2023, 25: 11055-11074.
7. Rácz G, Kerker S, Kovács Z, Vatai G, Ebrahimi M, Czermak P. Theoretical and experimental approaches of liquid entry pressure determination in membrane distillation processes. Periodica Polytechnica Chemical Engineering, 2014, 58(2): 81-91.
8. Silva RdS, Ramlow H, Santos BdC, Madalosso HB, Machado RAF, Marangoni C. Membrane distillation: Experimental evaluation of liquid entry pressure in commercial membranes with textile dye solutions. Journal of Water Process Engineering, 2021, 44: 102339.
9. Kaur R, Goyat R, Singh J, Umar A, Chaudhry V, Akbar S. An Overview of membrane distillation technology: One of the perfect fighters for desalination. Engineered Science, 2023, 21(771): 1-17.
10. Kalla S, Upadhyaya S, Singh K, Baghel R. Experimental and mathematical study of air gap membrane distillation for aqueous HCl azeotropic separation. Journal of Chemical Technology & Biotechnology, 2019, 94(1): 63-78.
11. Julian H, Nurgirisia N, Qiu G, Ting Y-P, Wenten IG. Membrane distillation for wastewater treatment: Current trends, challenges and prospects of dense membrane distillation. Journal of Water Process Engineering, 2022, 46: 102615.
12. Wu J, Zhang Y, Wang J, Zheng X, Chen Y. Municipal wastewater reclamation and reuse using membrane-based technologies: A review. Desalination and Water Treatment, 2021, 224: 65-82.

13. Xie Z, Hoang M, Duong T, Ng D, Singh P, Ray P, et al. Effect of membrane properties on performance of membrane distillation for ammonia removal. Journal of Materials Science Research, 2012, 1(1): 37-44.
14. Basma NS, Headen TF, Shaffer MSP, Skipper NT, Howard CA. Local structure and polar order in liquid N-Methyl-2-pyrrolidone (NMP). The Journal of Physical Chemistry B, 2018, 122(38): 8963-8971.
15. Tewodros BN, Yang DR, Park K. Design parameters of a direct contact membrane distillation and a case study of its applicability to low-grade waste energy. Membranes, 2022, 12(12).
16. Maqbool F, Soomro MI, Kumar L, Harijan K. Modeling and simulation of direct contact membrane distillation system integrated with a photovoltaic thermal for electricity and freshwater production. Frontiers in Energy Research, 2024, 12.
17. Liu Q, Zhao Y, Wu Y, Qin G, Li G, Lyu J. Impacts of salt concentration on nucleate pool boiling of NaCl solution. AIP Advances, 2023, 13(3): 035005.
18. Alsaadi AS, Francis L, Amy GL, Ghaffour N. Experimental and theoretical analyses of temperature polarization effect in vacuum membrane distillation. Journal of Membrane Science, 2014, 471: 138-148.
19. Hijaz HA, Zargar M, Shafieian A, Razmjou A, Khiadani M. Beyond the surface: Understanding temperature polarisation in DCMD membranes through varied operational parameters. Separation and Purification Technology, 2024, 350: 127978.
20. Karanikola V, Moore SE, Deshmukh A, Arnold RG, Elimelech M, Sáez AE. Economic performance of membrane distillation configurations in optimal solar thermal desalination systems. Desalination, 2019, 472: 114164.
21. James B, Colella W, Moton J. PEM Electrolysis H2A Production Case Study Documentation. National Renewable Energy Laboratory (NREL), 2013.
22. Harrison KW, Remick R, Martin GD, Hoskin A. Hydrogen production: Fundamentals and case study summaries. 18th World Hydrogen Energy Conference, May 16–21, 2010, Essen, Germany 2010.
23. Shiva Kumar S, Himabindu V. Hydrogen production by PEM water electrolysis – A review. Materials Science for Energy Technologies, 2019, 2(3): 442-4454.
24. Othman NH, Kabay N, Guler E. Principles of reverse electrodialysis and development of integrated-based system for power generation and water treatment: A review. Reviews in Chemical Engineering, 2022, 38(8): 921-958.
25. Veerman J, Vermaas DA. 4 - Reverse electrodialysis: Fundamentals. pp. 77-133. *In:* Cipollina A, Micale G (eds). Sustainable Energy from Salinity Gradients. Woodhead Publishing, 2016.
26. Energy Generation by Salinity Gradient Power-Reverse Electrodialysis. Italy 2021 [Available from: https://www.elgalabwater.com/blog/energy-generation-salinity-gradient-power-reverse-electrodialysis-italy
27. Hine F, Murakami K. Bubble effects on the solution IR drop in a vertical electrolyzer under free and forced convection. Journal of the Electrochemical Society, 1980, 127(2): 292-297.
28. Handojo L, Wardani AK, Regina D, Bella C, Kresnowati MTAP, Wenten IG. Electro-membrane processes for organic acid recovery. RSC Advances, 2019, 9(14): 7854-7869.
29. Khan ZU, Moronshing M, Shestakova M, Al-Othman A, Sillanpää M, Zhan Z, et al. Electro-deionization (EDI) technology for enhanced water treatment and desalination: A review. Desalination, 2023, 548: 116254.

30. Titorova VD, Moroz IA, Mareev SA, Pismenskaya ND, Sabbatovskii KG, Wang Y, et al. How bulk and surface properties of sulfonated cation-exchange membranes response to their exposure to electric current during electrodialysis of a Ca^{2+} containing solution. Journal of Membrane Science, 2022, 644: 120149.
31. Chaabouni A, Guesmi F, Louati I, Hannachi C, Hamrouni B. Temperature effect on ion exchange equilibrium between CMX membrane and electrolytes solutions. Journal of Water Reuse and Desalination. 2015, 5(4): 535-541.
32. Sarapulova V, Pismenskaya N, Butylskii D, Titorova V, Wang Y, Xu T, et al. Transport and electrochemical characteristics of CJMCED homogeneous cation exchange membranes in sodium chloride, calcium chloride, and sodium sulfate solutions. Membranes, 2020, 10(8): 165.
33. Ariono D, Khoiruddin, Subagjo, Wenten IG. Heterogeneous structure and its effect on properties and electrochemical behavior of ion-exchange membrane. Materials Research Express, 2017, 4(2): 024006.
34. Hansima MACK, Makehelwala M, Jinadasa KBSN, Wei Y, Nanayakkara KGN, Herath AC, et al. Fouling of ion exchange membranes used in the electrodialysis reversal advanced water treatment: A review. Chemosphere, 2021, 263: 127951.
35. Hansima MACK, Ketharani J, Samarajeewa DR, Nanayakkara KGN, Herath AC, Makehelwala M, et al. Probing fouling mechanism of anion exchange membranes used in electrodialysis self-reversible treatment by humic acid and calcium ions. Chemical Engineering Journal Advances, 2021, 8: 100173.
36. Myint MT, Ghassemi A, Nirmalakhandan N. Modeling in desalination—Electrodialysis reversal. Desalination and Water Treatment, 2011, 27(1): 255-267.
37. Venzke CD, Giacobbo A, Ferreira JZ, Bernardes AM, Rodrigues MAS. Increasing water recovery rate of membrane hybrid process on the petrochemical wastewater treatment. Process Safety and Environmental Protection, 2018, 117: 152-158.
38. Venzke CD, Gıacobbo A, Bernardes AM, Rodrıgues MAS. Petrochemical Industry: Wastewater treatment for water reuse. 15th International Conference on Environmental Science and Technology. 31st August-2nd September, Rhodes, Greece, 2017.
39. Ataei Far H, Hassani AH, Taghavi L, Fazeli M, Rashidi Mehrabadi A. Electro dialysis reversal (EDR) performance for reject brine treatment of reverse osmosis desalination system. PLOS ONE, 2022, 17(8): e0273240.
40. Turek M, Was J, Dydo P. Brackish water desalination in RO – Single pass EDR system. Desalination and Water Treatment, 2009, 7(1-3): 263-266.
41. Mitchell JK. On the penetration of gases. The American Journal of the Medical Sciences (1827-1924), 1833, 13(25): 100.
42. Fane AG, Wang R, Jia Y. Membrane technology: Past, present and future. Membrane and Desalination Technologies, 2011, 1-45.
43. Graham T. On the absorption and dialytic separation of gases by colloid septa. Part I: Action of a septum of caoutchouc. Journal of Membrane Science, 1995, 1(100): 27-31.
44. Ismail AF, David LIB. A review on the latest development of carbon membranes for gas separation. Journal of Membrane Science, 2001, 193(1): 1-18.
45. Stern SA. Polymers for gas separations: The next decade. Journal of Membrane Science, 1994, 94(1): 1-65.
46. Henis JMS, Tripodi MK. Composite hollow fiber membranes for gas separation: The resistance model approach. Journal of Membrane Science, 1981, 8(3): 233-246.

47. Kesting RE, Fritzsche AK. Polymeric gas separation membranes. New York: Wiley. 1993.
48. Kesting RE, Fritzsche AK, Murphy MK, Cruse CA, Handermann AC, Malon RF, et al. The second-generation polysulfone gas-separation membrane. I. The use of lewis acid: Base complexes as transient templates to increase free volume. Journal of Applied Polymer Science, 1990, 40(9-10): 1557-1574.
49. Pinnau I, Koros WJ. Relationship between substructure resistance and gas separation properties of defect-free integrally skinned asymmetric membranes. Industrial & Engineering Chemistry Research, 1991, 30(8): 1837-1840.
50. Kahlenberg L. On the nature of the process of osmosis and osmotic pressure with observations concerning dialysis. The Journal of Physical Chemistry, 2002, 10(3): 141-209.
51. Farber L. Applications of pervaporation. Science, 1935, 82(2120): 158-158.
52. Melnick JL. Concentration of dilute solutions of virus of mouse encephalo-myelitis by pervaporation and ultracentrifugation. Proceedings of the Society for Experimental Biology and Medicine, 1942, 49(4): 553-557.
53. Binning RC, Stuckey JM. Inventors method of separating hydrocarbons using ethyl cellulose permselective membrane patent US-2958657-A. 1960.
54. Binning RC, James FE. Permeation: A new commercial separation tool. Petrolum Refiner, 1958, 39(214): 88.
55. Binning RC, James FE. Now separate by membrane permeation. pet Refinery, 1958, 37(5): 214-215.
56. Binning RC, James FE. Permeation: A new way to separate mixtures. Oil Gas J. 1958, 56(21): 104-105.
57. Binning RC, Jennings JF, Martin EC. Inventors process for removing water from organic chemicals patent US5104545A. 1962.
58. Verkerk AW, Van Male P, Vorstman MAG, Keurentjes JTF. Description of dehydration performance of amorphous silica pervaporation membranes. Journal of Membrane Science, 2001, 193(2): 227-238.
59. Kirk RE, Othmer DF. Kirk-Othmer Encyclopedia of Chemical Technology. Wiley, 1978.
60. Aptel P, Cuny J, Jozefowicz J, Morel G, Neel J. Liquid transport through membranes prepared by grafting of polar monomers onto poly (tetrafluoroethylene) films. I. Some fractionations of liquid mixtures by pervaporation. Journal of Applied Polymer Science, 1972, 16(5): 1061-1076.
61. Aptel P, Challard N, Cuny J, Neel J. Application of the pervaporation process to separate azeotropic mixtures. Journal of Membrane Science, 1976, 1: 271-287.
62. Buonomenna MG. Membrane processes for a sustainable industrial growth. RSC Advances, 2013, 3(17): 5694-5740.
63. Scott K. Handbook of Industrial Membranes. Elsevier; 1995.
64. Sano T, Yanagishita H, Kiyozumi Y, Mizukami F, Haraya K. Separation of ethanol/water mixture by silicalite membrane on pervaporation. Journal of Membrane Science, 1994, 95(3): 221-228.
65. Feng X, Huang RYM. Liquid separation by membrane pervaporation: A review. Industrial & Engineering Chemistry Research, 1997, 36(4): 1048-1066.
66. Kita H, Horii K, Ohtoshi Y, Tanaka K, Okamoto K.-I. Synthesis of a zeolite NaA membrane for pervaporation of water/organic liquid mixtures. Journal of Materials Science Letters, 1995, 14(3): 206-208.

67. Kondo M, Komori M, Kita H, Okamoto K.-I. Tubular-type pervaporation module with zeolite NaA membrane. Journal of Membrane Science, 1997, 133(1): 133-141.
68. Lipnizki F, Field RW, Ten P.-K. Pervaporation-based hybrid process: A review of process design, applications and economics. Journal of Membrane Science, 1999, 153(2): 183-210.
69. Pivovar BS, Wang Y, Cussler EL. Pervaporation membranes in direct methanol fuel cells. Journal of Membrane Science, 1999, 154(2): 155-162.
70. Sunitha K, Rani KY, Moulik S, Satyanarayana SV, Sridhar S. Separation of NMP/water mixtures by nanocomposite PEBA membrane: Part I. Membrane synthesis, characterization and pervaporation performance. Desalination, 2013, 330: 1-8.
71. Kalyani S, Smitha B, Sridhar S, Krishnaiah A. Pervaporation separation of ethanol–water mixtures through sodium alginate membranes. Desalination, 2008, 229(1-3): 68-81.
72. Biduru S, Sridhar S, Suryanarayana Murthy G, Mayor S. Pervaporation of tertiary butanol/water mixtures through chitosan membranes cross-linked with toluylene diisocyanate. Journal of Chemical Technology & Biotechnology: International Research in Process, Environmental & Clean Technology, 2005, 80(12): 1416-1424.
73. Sridhar S, Smitha B, Reddy AA. Separation of 2-butanol-water mixtures by pervaporation through PVA–NYL 66 blend membranes. Colloids and Surfaces A: Physicochemical and Engineering Aspects, 2006, 280(1-3): 95-102.
74. Villaluenga JPG, Tabe-Mohammadi A. A review on the separation of benzene/cyclohexane mixtures by pervaporation processes. Journal of Membrane Science, 2000, 169(2): 159-174.
75. Lin L, Zhang Y, Kong Y. Recent advances in sulfur removal from gasoline by pervaporation. Fuel, 2009, 88(10): 1799-1809.
76. Peng M, Vane LM, Liu SX. Recent advances in VOCs removal from water by pervaporation. Journal of Hazardous Materials, 2003, 98(1-3): 69-90.
77. Qureshi N, Meagher MM, Huang J, Hutkins RW. Acetone butanol ethanol (ABE) recovery by pervaporation using silicalite–silicone composite membrane from fed-batch reactor of Clostridium acetobutylicum. Journal of Membrane Science, 2001, 187(1-2): 93-102.
78. O'Brien DJ, Senske GE, Kurantz MJ, Craig Jr JC. Ethanol recovery from corn fiber hydrolysate fermentations by pervaporation. Bioresource Technology, 2004, 92(1): 15-19.
79. Jain A, Jain A, Upadhyaya S, Dalai AK, Dalai AK, Chaurasia SP, et al. Pervaporation for ethanol-water separation and effect of fermentation inhibitors. Membrane processes: Pervaporation, vapor permeation and membrane distillation for industrial scale separations. 2018, 89-122.
80. Xue C, Liu F, Xu M, Zhao J, Chen L, Ren J, et al. A novel in situ gas stripping-pervaporation process integrated with acetone-butanol-ethanol fermentation for hyper n-butanol production. Biotechnology and Bioengineering, 2016, 113(1): 120-129.
81. Li L, Hou J, Ye Y, Mansouri J, Zhang Y, Chen V. Suppressing salt transport through composite pervaporation membranes for brine desalination. Applied Sciences, 2017, 7(8): 856.
82. Vallieres C, Favre E. Vacuum versus sweeping gas operation for binary mixtures separation by dense membrane processes. Journal of Membrane Science, 2004, 244(1): 17-23.

83. Liu G, Jin W. Pervaporation membrane materials: Recent trends and perspectives. Journal of Membrane Science, 2021, 636: 119557.
84. Ahmad AA-O, Shafie ZA-O, Zaulkiflee NA-O, Pang WA-O. Preliminary study of emulsion liquid membrane formulation on acetaminophen removal from the aqueous phase. Membranes, 9: 133.
85. Mohammed AA, Al-Khateeb RW. Application of emulsion liquid membrane using green surfactant for removing phenol from aqueous solution: Extraction, stability and breakage studies. Journal of Ecological Engineering, 2022, 23(1): 305-314.
86. Huang C-R, Fan H, Zhou D. A closed-form solution for a mathematical model of emulsion liquid membrane. Journal of Membrane Science, 2009, 339(1): 233-238.
87. Kowalczuk PB, Drzymala J. Physical meaning of the Sauter mean diameter of spherical particulate matter. Particulate Science and Technology, 2016, 34(6): 645-647.
88. Saik Su G, Morad N, Ismail N, Rafatullah M. Developments in supported liquid membranes for treatment of metal-bearing wastewater. Separation & Purification Reviews, 2022, 51(1): 38-56.
89. Farah M, Giralt J, Stüber F, Font J, Fabregat A, Fortuny A. Supported liquid membranes for the removal of pharmaceuticals from aqueous solutions. Journal of Water Process Engineering, 2022, 49: 103170.
90. Alemrajabi M, Ricknell J, Samak S, Rodriguez Varela R, Martinez J, Hedman F, et al. Separation of rare-earth elements using supported liquid membrane extraction in pilot scale. Industrial & Engineering Chemistry Research, 2022, 61(50): 18475-18491.
91. Ata ON. Mathematical modelling of unsteady-state transport of metal ions through supported liquid membrane. Hydrometallurgy, 2007, 87(3): 148-156.
92. Szpakowska M, Nagy OB. Non-steady state vs. steady state kinetic analysis of coupled ion transport through binary liquid membranes. Journal of Membrane Science, 1993, 76(1): 27-38.
93. Ocaña-González JA, Aranda-Merino N, Pérez-Bernal JL, Ramos-Payán M. Solid supports and supported liquid membranes for different liquid phase microextraction and electromembrane extraction configurations. A review. Journal of Chromatography A, 2023, 1691: 463825.
94. Mubashir M, D'Angelo FN, Gallucci F. Recent advances and challenges of deep eutectic solvent based supported liquid membranes. Separation & Purification Reviews, 2022, 51(2): 226-244.
95. Söldner A, Zach J, König B. Deep eutectic solvents as extraction media for metal salts and oxides exemplarily shown for phosphates from incinerated sewage sludge ash. Green Chemistry, 2019, 21(2): 321-328.
96. Kislik VS. Bulk liquid membrane. pp. 1-2. *In:* Drioli E, Giorno L (eds). Encyclopedia of Membranes. Berlin, Heidelberg: Springer Berlin Heidelberg, 2015.
97. Bartsch RA, Way JD. Chemical separations with liquid membranes: An overview. ACS Symposium Series. 642: American Chemical Society, 1996, 1-10.
98. Chang SH. Types of bulk liquid membrane and its membrane resistance in heavy metal removal and recovery from wastewater. Desalination and Water Treatment, 2016, 57(42): 19785-19793.
99. Bartsch RA, Way JD. Chemical separations with liquid membranes: An overview. pp. 1-10. *In:* Bartsch RA, Way JD (eds). Chemical Separations with Liquid Membranes. 642: American Chemical Society, 1996.

100. Memon FN, Memon S, Minhas FT. Rapid transfer of methyl red using calix[6] arene as a carrier in a bulk liquid membrane. Comptes Rendus Chimie, 2014, 17(6): 577-585.
101. Wahab HS, Albasri SAMM. Removal of acetaminophen residues from wastewater by bulk liquid membrane process. Iraqi Journal of Chemical and Petroleum Engineering, 2020, 21(4): 1-9.
102. Huang D-S, Yi Z-Z, Huang Z-L, Yi P, Li Z-J, Liu W. Mass transfer mechanism and mathematical model for extraction process of L-theanine across bulk liquid membrane. Iranian Journal of Chemistry and Chemical Engineering, 2012, 31(2): 53-58.
103. Candela AM, Benatti V, Palet C. Pre-concentration of uranium (VI) using bulk liquid and supported liquid membrane systems optimized containing bis(2-ethylhexyl) phosphoric acid as carrier in low concentrations. Separation and Purification Technology, 2013, 120: 172-179.
104. Pandey K. Groundwater in 12 Indian states found to be contaminated with uranium: DownToEarth 2023 [Available from: https://www.downtoearth.org.in/water/groundwater-in-12-indian-states-found-to-be-contaminated-with-uranium-87096.
105. Tillman FD, Beisner KR, Anderson JR, Unema JA. An assessment of uranium in groundwater in the Grand Canyon region. Scientific Reports, 2021, 11(1): 22157.
106. Keep the Canyon Grand 2024 [Available from: https://www.grandcanyontrust.org/keep-the-canyon-grand.
107. Pirmoradi M, Ashrafizadeh SN. Removal of nitrate from water by bulk liquid membrane. Desalination and Water Treatment, 2017, 58: 228–238.
108. Rounaghi GH, Ghaemi A, Chamsaz M. Separation study of some heavy metal cations through a bulk liquid membrane containing 1,13-bis(8-quinolyl)-1,4,7,10,13-pentaoxatridecane. Arabian Journal of Chemistry, 2016, 9: S490- S496.
109. Rounaghi GH, Kakhki RMZ, Chamsaz M. Transport study of some transition metal cations through a bulk liquid membrane using dicyclohexyl-18-crown-6 as carrier. Asian Journal of Chemistry, 2011, 23(3): 1167-1172.
110. Kolff WJ, Watschinger B. Further development of a coil kidney; disposable artificial kidney. The Journal of Laboratory and Clinical Medicine, 1956, 47(6): 969-977.
111. Mitsides N, Keane DF, Lindley E, Mitra S. Technology innovation for patients with kidney disease. Journal of Medical Engineering and Technology, 2014, 39(7): 424-433.
112. Brescia MJ, Cimino JE, Appel K, Hurwich BJ. Chronic hemodialysis using venipuncture and a surgically created arteriovenous fistula. The New England Journal of Medicine, 1966, 275(20): 1089-1092.
113. Lytton B, Goffinet JA, May CJ, Weiss RM. Experience with arteriovenous fistula in chronic hemodialysis. The Journal of Urology, 1970, 104(4): 512-517.
114. Mumtaz A, Anees M, Bilal M, Ibrahim M. Beta-2 microglobulin levels in hemodialysis patients. Saudi J Kidney Dis Transpl, 2010, 21(4): 701-706.
115. Thomas M, Moriyama K, Ledebo I. AN69: Evolution of the world's first high permeability membrane. Contributions to Nephrology, 2011, 173: 119-129.
116. Laude-Sharp M, Caroff M, Simard L, Pusineri C, Kazatchkine MD, Haeffner-Cavaillon N. Induction of IL-1 during hemodialysis: Transmembrane passage of intact endotoxins (LPS). Kidney International, 1990, 38(6): 1089-1094.
117. Dees K. Dialysis, Types 2023 [Available from: https://www.cincinnatichildrens.org/health/d/dialysis-type.
118. Craddock PR, Fehr J, Dalmasso AP, Brighan KL, Jacob HS. Hemodialysis

leukopenia. Pulmonary vascular leukostasis resulting from complement activation by dialyzer cellophane membranes. The Journal of Clinical Investigation, 1977, 59(5): 879–888.

119. Yamashita AC, Tomisawa N, Takezawa A, Sakurai K, Sakai T. Blood compatibility and filtration characteristics of a newly developed polyester polymer alloy membrane. Haemodialysis International, 2004, 8(4): 368-371.
120. Peek GJ, Clemens F, Elbourne D, Firmin R, Hardy P, Hibbert C, et al. CESAR: Conventional ventilatory support vs extracorporeal membrane oxygenation for severe adult respiratory failure. BMC Health Services Research, 2006, 6(1): 163.
121. Huang J, Peng L, Zeng G, Li X, Zhao Y, Liu L, et al. Evaluation of micellar enhanced ultrafiltration for removing methylene blue and cadmium ion simultaneously with mixed surfactants. Separation and Purification Technology, 2014, 125: 83-89.
122. Verma SP, Sarkar B. Rhamnolipid based micellar-enhanced ultrafiltration for simultaneous removal of Cd(II) and phenolic compound from wastewater. Chemical Engineering Journal, 2017, 319: 131-142.
123. Verma SP, Sarkar B. Simultaneous removal of Cd (II) and p-cresol from wastewater by micellar-enhanced ultrafiltration using rhamnolipid: Flux decline, adsorption kinetics and isotherm studies. Journal of Environmental Management, 2018, 213: 217-235.
124. Zeng G-M, Li X, Huang J-H, Zhang C, Zhou C-F, Niu J, et al. Micellar-enhanced ultrafiltration of cadmium and methylene blue in synthetic wastewater using SDS. Journal of Hazardous Materials, 2011, 185(2): 1304-1310.
125. El-Abbassi A, Khayet M, Hafidi A. Micellar enhanced ultrafiltration process for the treatment of olive mill wastewater. Water Research, 2011, 45(15): 4522-4530.
126. Häyrynen P, Landaburu-Aguirre J, Pongrácz E, Keiski RL. Study of permeate flux in micellar-enhanced ultrafiltration on a semi-pilot scale: Simultaneous removal of heavy metals from phosphorous rich real wastewaters. Separation and Purification Technology, 2012, 93: 59-66.
127. Namaghi HA, Mousavi SM. Micellar-enhanced ultrafiltration of soft drink wastewater using anionic and mixed anionic/nonionic surfactants. Journal of the Taiwan Institute of Chemical Engineers, 2014, 45(4): 1850-1854.
128. Chang W-S, Chen S-S, Sie C-H, Nguyen NC, Cheng H-H, Hsu H-T. Recovery of chromium from plastic plating wastewater by cetyltrimethylammonium bromide MEUF and electrodialysis. Desalination and Water Treatment, 2015, 55(9): 2408-2415.
129. Aoudia M, Allal N, Djennet A, Toumi L. Dynamic micellar enhanced ultrafiltration: Use of anionic (SDS)–nonionic(NPE) system to remove Cr^{3+} at low surfactant concentration. Journal of Membrane Science, 2003, 217(1): 181-192.
130. Fernández E, Benito JM, Pazos C, Coca J. Ceramic membrane ultrafiltration of anionic and nonionic surfactant solutions. Journal of Membrane Science, 2005, 246(1): 1-6.
131. Byhlin H, Jönsson AS. Influence of adsorption and concentration polarisation on membrane performance during ultrafiltration of a non-ionic surfactant. Desalination, 2003, 151(1): 21-31.
132. Jung J, Yang J-S, Kim S-H, Yang J-W. Feasibility of micellar-enhanced ultrafiltration (MEUF) or the heavy metal removal in soil washing effluent. Desalination, 2008, 222(1): 202-211.
133. Sabaté J, Pujolà M, Centelles E, Galán M, Llorens J. Determination of equilibrium distribution constants of phenol between surfactant micelles and water using

ultrafiltering centrifuge tubes. Colloids and Surfaces A: Physicochemical and Engineering Aspects, 1999, 150(1): 229-245.
134. Brant LL, Stellner KL, Scamehorn JF. Recovery of surfactant from surfactant-based separations using a precipitation process. pp. 323-338. *In:* Surfactant-based Separation Processes, CRC Press, 2020.
135. Juang R-S, Xu Y-Y, Chen C-L. Separation and removal of metal ions from dilute solutions using micellar-enhanced ultrafiltration. Journal of Membrane Science, 2003, 218(1-2): 257-267.
136. Qu Y-H, Zeng G-M, Huang J-H, Xu K, Fang Y-Y, Li X, et al. Recovery of surfactant SDS and Cd^{2+} from permeate in MEUF using a continuous foam fractionator. Journal of Hazardous Materials, 2008, 155(1-2): 32-38.
137. Li S, Pi Y, Bao M, Zhang C, Zhao D, Li Y, et al. Effect of rhamnolipid biosurfactant on solubilization of polycyclic aromatic hydrocarbons. Marine Pollution Bulletin, 2015, 101(1): 219-225.
138. Purkait MK, DasGupta S, De S. Removal of dye from wastewater using micellar-enhanced ultrafiltration and recovery of surfactant. Separation and Purification Technology, 2004, 37(1): 81-92.
139. Verma SP, Sarkar B. Use of rhamnolipid in micellar-enhanced ultrafiltration for simultaneous removal of Cd^{2+} and crystal violet from aqueous solution. Asia-Pacific Journal of Chemical Engineering, 2019, 14(3): e2315.
140. Samal K, Das C, Mohanty K. Eco-friendly biosurfactant saponin for the solubilization of cationic and anionic dyes in aqueous system. Dyes and Pigments, 2017, 140: 100-108.
141. Dwars T, Paetzold E, Oehme G. Reactions in micellar systems. Angewandte Chemie International Edition, 2005, 44(44): 7174-7199.
142. La Sorella G, Strukul G, Scarso A. Recent advances in catalysis in micellar media. Green Chemistry, 2015, 17(2): 644-683.
143. Gallou F, Isley NA, Ganic A, Onken U, Parmentier M. Surfactant technology applied toward an active pharmaceutical ingredient: More than a simple green chemistry advance. Green Chemistry, 2016, 18(1): 14-19.
144. Kulkarni T, Slaughter G. Application of semipermeable membranes in glucose biosensing. Membranes, 2016, 6(4): 55-75.
145. Mross S, Fürst P, Pierrat S, Zimmermann T, Vogt H, Kraft M. Enzyme sensor with polydimethylsiloxane membrane and CMOS potentiostat for wide-range glucose measurements. IEEE Sensors Journal, 2015, 15(12): 7096-7104.
146. Luo T, Wang H, Song H, Christen JB (eds). CMOS potentiostat for chemical sensing applications. SENSORS, 2013 IEEE, 2013 4-6 Nov., Baltimore, Maryland, USA.
147. Zhao H, Ju H. Multilayer membranes for glucose biosensing via layer-by-layer assembly of multiwall carbon nanotubes and glucose oxidase. Analytical Biochemistry, 2006, 350(1): 138-144.
148. Nikoleli G-P, Israr MQ, Tzamtzis N, Nikolelis DP, Willander M, Psaroudakis N. Structural characterization of graphene nanosheets for miniaturization of potentiometric urea lipid film based biosensors. Electroanalysis, 2012, 24(6): 1285-1295.
149. Rossokhaty V, Rossokhata N. Mathematical model of a biosensor with multilayer charged membrane. Computer Physics Communications, 2002, 147(1): 366-369.
150. Wang Q, Jiao C, Wang X, Wang Y, Sun K, Li L, et al. A hydrogel-based biosensor for stable detection of glucose. Biosensors and Bioelectronics, 2023, 221: 114908.

151. Zhenglan B, Yue Q, Liang X, Anduo H, Hui Y, Fenghong C. Glucose biosensing based on a hydrogel optical fiber immobilized with glucose oxidase. Optik, 2022, 255: 168655.
152. Jin R, Wang F, Li Q, Yan X, Liu M, Chen Y, et al. Construction of multienzyme-hydrogel sensor with smartphone detector for on-site monitoring of organophosphorus pesticide. Sensors and Actuators B: Chemical, 2021, 327: 128922.
153. Chen M, Wang Y, Zhang J, Peng Y, Li S, Han D, et al. Stimuli-responsive DNA-based hydrogels for biosensing applications. Journal of Nanobiotechnology, 2022, 20(1): 40.
154. Shahbazi M-A, Bauleth-Ramos T, Santos HA. DNA Hydrogel assemblies: Bridging synthesis principles to biomedical applications. Advanced Therapeutics, 2018, 1(4): 1800042.
155. Setapa A, Ahmad N, Mohd Mahali S, Mohd Amin MCI. Mathematical model for estimating parameters of swelling drug delivery devices in a two-phase release. Polymers, 2020, 12(12): 2921.
156. Drury JL, Mooney DJ. Hydrogels for tissue engineering: Scaffold design variables and applications. Biomaterials, 2003, 24(24): 4337-4351.
157. Pan J-f, Yuan H-F, Guo C-A, Liu J, Geng X-H, Fei T, et al. One-step cross-linked injectable hydrogels with tunable properties for space-filling scaffolds in tissue engineering. RSC Advances, 2015, 5(51): 40820-40830.
158. Liu J, Yang L, Liu K, Gao F. Hydrogel scaffolds in bone regeneration: Their promising roles in angiogenesis. Frontiers in Pharmacology, 2023, 14.
159. Sivaraj D, Chen K, Chattopadhyay A, Henn D, Wu W, Noishiki C, et al. Hydrogel scaffolds to deliver cell therapies for wound healing. Frontiers in Bioengineering and Biotechnology, 2021, 9.
160. Lee MJ, Espinosa-Marzal RM. Intrinsic and extrinsic tunability of double-network hydrogel strength and lubricity. ACS Applied Materials & Interfaces, 2023, 15(16): 20495-20507.
161. Lane T, Holloway JL, Milani AH, Saunders JM, Freemont AJ, Saunders BR. Double network hydrogels prepared from pH-responsive doubly crosslinked microgels. Soft Matter, 2013, 9(33): 7934-7941.
162. Zhou Y, Pan M, Lin R, Huang H. Advances in hydrogel materials applied to pancreatic-related diseases. Journal of Pancreatology, 2024, 7(3).
163. Huan Z, Li J, Guo J, Yu Y, Li L. Pancreatic islet cells in microfluidic-spun hydrogel microfibers for the treatment of diabetes. Acta Biomaterialia, 2024, 187: 149-160.
164. Xu L, Ji H, Zhong R, Cheng S, Dang G, Xu T, et al. Antioxidative hydrogel-embedded polyethersulfone membrane with improved hemocompatibility to alleviate oxidative stress. Journal of Membrane Science, 2023, 684: 121866.
165. Sarkar D, Datta D, Sen D, Bhattacharjee C. Simulation of continuous stirred rotating disk-membrane module: An approach based on surface renewal theory. Chemical Engineering Science, 2011, 66(12): 2554-2567.
166. Niemi H, Bulsari A, Palosaari S. Simulation of membrane separation by neural networks. Journal of Membrane Science, 1995, 102: 185-191.
167. Curcio S, Calabrò V, Iorio G. Reduction and control of flux decline in cross-flow membrane processes modeled by artificial neural networks and hybrid systems. Desalination, 2009, 236(1): 234-243.
168. Bhattacharjee C, Sen D, Sarkar P, Datta S, Bhattacharya PK. Studies on the application of different ANNs to predict permeate flux in rotating disk membrane

modules: A case study with MATLAB™. Desalination and Water Treatment, 2009, 2: 170-184.
169. Sen D, Roy A, Bhattacharya A, Banerjee D, Bhattacharjee C. Development of a knowledge based hybrid neural network (KBHNN) for studying the effect of diafiltration during ultrafiltration of whey. Desalination, 2011, 273(1): 168-178.
170. Sargolzaei J, Haghighi Asl M, Hedayati Moghaddam A. Membrane permeate flux and rejection factor prediction using intelligent systems. Desalination, 2012, 284: 92-99.
171. Sen D, Roy W, Das L, Sadhu S, Bhattacharjee C. Ultrafiltration of macromolecules using rotating disc membrane module (RDMM) equipped with vanes: Effects of turbulence promoter. Journal of Membrane Science, 2010, 360(1): 40-47.
172. Jang JSR. ANFIS: Adaptive-network-based fuzzy inference system. IEEE Transactions on Systems, Man, and Cybernetics, 1993, 23(3): 665-685.
173. Ion I. A mamdani type fuzzy logic controller. pp. 325-350. *In:* Elmer PD (ed.). Fuzzy Logic – Controls, Concepts, Theories and Applications. InTech, 2012.
174. Darvishmanesh S, Qian X, Wickramasinghe SR. Responsive membranes for advanced separations. Current Opinion in Chemical Engineering, 2015, 8: 98-104.
175. Himstedt HH, Sengupta A, Qian X, Wickramasinghe SR. Magnetically responsive nano filtration membranes for treatment of coal bed methane produced water. Journal of the Taiwan Institute of Chemical Engineers, 2019, 94: 97-108.
176. Himstedt HH, Yang Q, Dasi LP, Qian X, Wickramasinghe SR, Ulbricht M. Magnetically activated micromixers for separation membranes. Langmuir, 2011, 27(9): 5574-5581.
177. Dutta N, Umashankar S, Arun Shanker VK, Padmanaban S, Leonowicz Z, Wheeler P. Centrifugal pump cavitation detection using machine learning algorithm technique. International Conference on Environment and Electrical Engineering (EEEIC), 12-15th June. Palermo, Italy. IEEE, 2018.
178. Lin H. Types of Machine Learning Algorithm 2017 [Available from: https://scientistcafe.com/2017/07/08/machinelearningal.
179. Xue H, Yang Q, Chen S. SVM: Support vector machines. pp. 36-60. *In:* Wu X, Kumar V (eds). The Top Ten Algorithms in Data Mining. CRC Press, 2009.
180. Bhattacharjee C, Sen D, Sarkar P, Datta S, Bhattacharya PK. Studies on the application of different ANNs to predict permeate flux in rotating disk membrane modules: A case study with MATLAB™. Desalination and Water Treatment, 2009, 2(1): 170-183.
181. Odabaşı Ç, Dologlu P, Gülmez F, Kuşoğlu G, Çağlar Ö. Investigation of the factors affecting reverse osmosis membrane performance using machine-learning techniques. Computers & Chemical Engineering, 2022, 159: 107669.
182. Waqas S, Harun NY, Sambudi NS, Arshad U, Nordin NA, Bilad MR, et al. SVM and ANN modelling approach for the optimization of membrane permeability of a membrane rotating biological contactor for wastewater treatment. Membranes, 2022, 12(9): 821.
183. Antonelli G, Filippi J, D'Orazio M, Curci G, Casti P, Mencattini A, et al. Integrating machine learning and biosensors in microfluidic devices: A review. Biosensors and Bioelectronics, 2024, 263: 116632.
184. Gonzalez-Navarro FF, Stilianova-Stoytcheva M, Renteria-Gutierrez L, Belanche-Muñoz LA, Flores-Rios BL, Ibarra-Esquer JE. Glucose oxidase biosensor modeling and predictors optimization by machine learning methods. Sensors, 2016, 16(11).

Index

T

U

V

W

Y

Z